Progress in Probability and Statistics
Volume 12

Peter Huber
Murray Rosenblatt
series editors

Seminar on Stochastic Processes, 1985

E. Çınlar
K.L. Chung
R.K. Getoor
editors

J. Glover
managing editor

1986 Springer Science+Business Media, LLC

E. Çınlar
Civil Engineering Department
Princeton University
Princeton, NJ 08544
U.S.A.

K.L. Chung
Department of Mathematics
Stanford University
Stanford, CA 94305
U.S.A.

R.K. Getoor
Department of Mathematics
University of California–San Diego
La Jolla, CA 92093
U.S.A.

J. Glover (managing editor)
Department of Mathematics
University of Florida
Gainesville, FL 32611
U.S.A.

Library of Congress Cataloging in Publication Data
Seminar on Stochastic Processes (5th: 1985: University of Florida)
Seminar on Stochastic Processes, 1985.
(Progress in probability and statistics; vol. 12)
Includes bibliographies.
1. Stochastic processes—Congresses. I. Çınlar,
E. (Erhan), 1941- . II. Chung, Kai Lai, 1917- .
III. Getoor, R.K. (Ronald Kay), 1929- .
IV. Title. V. Series: Progress in probability and statistics; v.12.
QA274.A1S45 1985 519.2 86-11285

CIP-Kurztitelaufnahme der Deutschen Bibliothek
Seminar on Stochastic Processes:
Seminar on Stochastic Processes . . . –Boston; Basel
Stuttgart : Birkhäuser
5, 1985 (1986).
(Progress in probability and statistics; Vol. 12)
ISBN 978-1-4684-6750-5 ISBN 978-1-4684-6748-2 (eBook)
DOI 10.1007/978-1-4684-6748-2

NE: GT

Originally published by Birkhäuser Boston, Inc. in 1986
Softcover reprint of the hardcover 1st edition 1986

FOREWORD

The 1985 Seminar on Stochastic Processes was held at the University of Florida, Gainesville, in March. It was the fifth seminar in a continuing series of meetings which provide opportunities for researchers to discuss current work in stochastic processes in an informal atmosphere. Previous seminars were held at Northwestern University, Evanston and the University of Florida, Gainesville.

The participants' enthusiasm and interest have resulted in stimulating and successful seminars. We thank them for it, and we also thank those participants who have permitted us to publish their research here.

The seminar was made possible through the generous supports of the Division of Sponsored Research and the Department of Mathematics of the University of Florida, and the Air Force Office of Scientific Research, Grant No. 82-0189. We are grateful for their support. Finally, the comfort and hospitality we enjoyed in Gainesville were due to the splendid efforts of Professor Zoran Pop-Stojanovic.

J. G.

Gainesville, 1986.

Table of Contents

Seminar on Stochastic Processes, 1985
Birkhäuser, Boston, 1986

A DECOMPOSITION OF EXCESSIVE MEASURES

by

R. M. Blumenthal

Let $\{P_t; t \geqslant 0\}$ denote the transition semigroup for a
Borel right Markov process on a state space $(E, \mathcal{E})$. We set
$U(x,D) = \int_0^\infty P_t(x,D)dt$ and denote by μU and Uf the potential
of a measure μ and a function f respectively. Given a
measure m on $\mathcal{E}$ and a set D in $\mathcal{E}$ let m_D be the measure
$m_D(B) = m(D \cap B)$. A measure m on $\mathcal{E}$ is called excessive if
m is σ-finite and $m \geqslant mP_t$ for all t. If μ is a measure on
$\mathcal{E}$ the formula $\mu U(A) = \int \mu(dx)U(x,A)$ defines a measure, the
potential of μ, on $\mathcal{E}$. It will be excessive if and only if
it is σ-finite. An excessive measure m is called invariant
if $mP_t = m$ for all t.

In [1] Fitzsimmons and Maisonneuve call an excessive
measure m dissipative if m is the set-wise supremum of the
potentials μU which are set-wise less than m, and they call
m conservative if the only potential less than m is the
zero measure. Then they prove that every excessive m has a
unique decomposition into a sum, $m^c + m^d$, of a conservative
and a dissipative measure. In fact they prove that if q is
any strictly positive Borel function on E with $m(q)$ finite

then m^c is m restricted to $\{U_q = \infty\}$ and m^d is m restricted to $\{U_q < \infty\}$ so that up to m-null sets these sets are independent of q.

The purpose of this note is to prove that for every excessive function ϕ and positive t we have $\phi = P_t\phi$ almost everywhere relative to m^c. It turns out that at the expense of a few extra lines one can obtain this invariance result and the $m^c + m^d$ decomposition all at once, and without the intervention of the Kuznecov theory of stationary measures that formed the basis for [1]; and so we will give this slightly expanded presentation. The invariance result verifies a conjecture of Getoor and Steffens, who came upon it while preparing their paper [3] on capacity. I appreciate their calling the problem to my attention.

We will start off with some definitions and simple observations.

Call a set A dissipative if A is in $\mathscr{E}$, A is finely open, and there is a function h in $\mathscr{E}^+$ with h strictly positive on A and $Uh \leq 1$ on all of E. One argues easily that any countable union of dissipative sets is itself dissipative. For an example of such a set take a bounded Borel excessive ϕ such that $P_t\phi$ decreases to zero as t increases to infinity, and set $A = \{\phi > P_r\phi\}$ where r is a positive number. Then $U(\phi - P_r\phi) = \int_0^r P_t\phi dt$ so we can take $h = (\phi - P_r\phi)/r\|\phi\|$. The requirement that $P_t\phi$ decreases to zero is unnecessary; we can replace any bounded ϕ by $\psi = \phi - \theta$, where θ is the limit as t approaches ∞ of $P_t\phi$. Then ψ is excessive, $P_t\psi$ decreases to zero and $\{\psi > P_r\psi\} = \{\phi > P_r\phi\}$. Also ϕ need not be bounded or even finite because we can replace ϕ by $\min(\phi,n)$ and use the fact that a countable union of dissipative sets is again one. Any set of the form $\{0 < Uq < \infty\}$ with q in $\mathscr{E}^+$ is a countable union of sets $\{Uq > P_r Uq\}$ and so is dissipative.

Now given an excessive measure m, take a finite measure θ equivalent to m and a sequence $\{A_n\}$ of dissipative sets such that $\lim_n \theta(A_n) = \sup\{\theta(D) | D$ dissipative$\}$ and let $\bar{A}$ be the union of the A_n. Then $\bar{A}$ is dissipative, and obviously if D is any dissipative set then $\theta(D - \bar{A})$ is zero so that $m(D - \bar{A})$ is zero also. In particular if ϕ is any Borel excessive function and t is positive we have $\phi(x) = P_t\phi(x)$ for almost all (m)x in $E - \bar{A}$. We want to replace $\bar{A}$ by a slightly larger set. Specifically, take a function h in $\mathscr{E}^+$ which is strictly positive on $\bar{A}$ and with $Uh < 1$ on E and set $A = \{Uh > 0\}$. Then A is in $\mathscr{E}$, A is dissipative and $\bar{A}$ is contained in A, so that $m(A - \bar{A}) = 0$. Let B be the complement of A. Since A is the set where an excessive function is strictly positive, B must be absorbing; that is $P^x(X_t$ is in A for some $t > 0) = 0$ for all x in B. The sets A and B are the ones referred to in the next statement.

THEOREM 1. m_A <u>is dissipative</u>. m_B <u>is conservative. If</u> ϕ <u>is excessive then for every</u> t, $\phi = P_t\phi$ <u>almost everywhere</u> m_B.

PROOF. Since A and $\bar{A}$ differ by an m-null set we have established already, in the previous paragraph, the assertion about ϕ, at least if ϕ is a Borel function. This implies the conclusion for a general excessive ϕ because according to (6.11) of Getoor and Sharpe [2] for any excessive ϕ there is a Borel excessive ψ with the equalities $\phi = \psi$ and $P_t\phi = P_t\psi$ holding almost surely (m).

Before continuing the proof we will make some remarks about measures.

(1.1) if M and N are σ-finite measures with $M(D) \geq N(D)$ for all D in $\mathscr{E}$ then there is a unique measure θ on $\mathscr{E}$ with $N + \theta = M$. Of course $\theta(f) = M(f) - N(f)$ if f is in $\mathscr{E}^+$ and M(f) is finite. We will just write M - N when we mean θ;

(1.2) if $M_1, M_2, \ldots$ is a sequence of σ-finite measures with $M_1(C) \geq M_2(C) \geq \ldots$ for all C in $\mathscr{E}$ then there is a unique measure ψ such that $\psi(C) = \lim_n M_n(C)$ whenever $M_n(C)$ is finite for some n. We write simply $\psi = \lim_n M_n$.

(1.3) suppose m is excessive, t_n is a sequence of numbers increasing to ∞ and we set $\psi = \lim_n mP_{t_n}$ in the interpretation from (1.2). Then ψ is independent of the sequence t_n, ψ is invariant and $\theta = m - \psi$ is excessive with $\lim_{t\to\infty} \theta P_t(D) = 0$ whenever $m(D) < \infty$.

Assertions (1.1) through (1.3) are trivial to verify.

Now let θ be excessive and suppose there is an increasing sequence $\{C_n\}$ of sets whose union is E and such that for each n, $\theta P_t(C_n) \to 0$ as $t \to \infty$. Then

(1.4) $$(\theta - \theta P_t)U(F) = \int_0^t \theta P_r(F)\,dr$$

for all F in $\mathscr{E}$. To see this note that (1.4) follows from the usual semigroup manipulation whenever $\lim_{R\to\infty} \int_R^{R+t} \theta P_s(F)ds = 0$, in particular for any F contained in some C_n. But each side of (1.4) is a measure in F so the equality holds for all Borel F.

To complete the proof of Theorem 1 we will verify the following assertions:

(1.5) m_B is excessive. This is an immediate consequence of the fact that m is excessive and B is absorbing.

(1.6) if $\mu U \leq m_B$ then $\mu U = 0$. If not then there is a set C in $\mathscr{E}$ with $0 < \mu U(C) < \infty$. Then $\mu P_t U(C)$ approaches 0 as t tends to ∞, and so for some t, $\mu(\{UI_C > P_t UI_C\})$ is strictly positive. Call the set in braces D. It is finely open and so $U(x,D) > 0$ for all x in D. But then $m_B(D) \geq \int \mu(dx)U(x,D) \geq \int_D \mu(dx)U(x,D) > 0$, in violation of the fact that D is dissipative.

(1.7) m_B is invariant. As in (1.3) write $m_B = \theta + \phi$ with ϕ invariant and θ satisfying $\theta P_t(F) \to 0$ as $t \to \infty$ whenever $m_B(F)$ is finite. By (1.4) $t^{-1}(\theta - \theta P_t)U = t^{-1}\int_0^t \theta P_s ds \leq m_B$, and so by (1.6), $t^{-1}\int_0^t \theta P_s ds$ is the zero measure. Since this increases to θ as t decreases to zero it follows that θ is zero. That is, $m_B = \phi$ and hence is invariant.

(1.8) m_A is excessive. This is immediate from the fact that $m = m_A + m_B$ and m_B is invariant.

(1.9) $m_A = \sup\{\mu U | \mu U \leq m_A\}$. To prove this take an increasing sequence $\{A_n\}$ of sets such that $m(A_n)$ is finite for each n and such that A is the union of the A_n. Let h in $\mathcal{E}^+$ be strictly positive on A and such that $Uh \leq 1$ on E. Fix n for a moment and set $\mu = nm_{A_n}$. Then μ is a finite measure. It is dominated by a multiple of m_A and so $\mu U(B) = 0$, m_A being excessive and carried by A. Also we have $\|\mu\| \geq \mu Uh \geq \varepsilon \mu U(\{h \geq \varepsilon\})$ and so writing $E = \cup_n \{h > 1/n\} \cup B$ we see that μU is σ-finite. Let ϕ_n be the excessive measure $nm_{A_n}U \wedge m_A$. It is dominated by a potential and so the calculation

$$n(\phi_n - \phi_n P_{1/n})U = n \int_0^{n^{-1}} \phi_n P_s ds$$

is valid. This sequence of potentials is increasing with n, it is dominated by m_A and its limit exceeds ϕ_k for each k. We will complete the proof by showing that the sequence increases to m_A. Let θ_n be the measure $\int_0^1 m_{A_n} P_s ds$, which is dominated by m_A. If f_n is the Radom-Nikodym derivative of θ_n relative to m_A then $f_1 \leq f_2 \leq \ldots \leq 1$, a.e. (m_A). The measures θ_n increase to the measure $\int_0^1 m_A P_s ds$, which is equivalent to m_A and so the limit of the f_n is strictly positive a.e. (m_A). The measure $n\theta_n \wedge m_A$ has derivative $nf_n \wedge 1$ relative to m_A and this sequence of functions increases to 1. The measure ϕ_n exceeds $n\theta_n \wedge m_A$ and so the desired conclusion follows easily.

This completes the proof of Theorem 1. Specifically (1.9) is the statement that m_A is dissipative, and (1.6), that m_B is conservative.

We should include a proof of the characterization in terms of finiteness of potentials.

(1.10) if $q \in \mathcal{E}^+$, $q > 0$ and $m(q) < \infty$ then almost surely (m) we have $A = \{Uq < \infty\}$ and $B = \{Uq = \infty\}$. The first assertion is obvious: specifically Uq is strictly positive since q is and so $\{Uq < \infty\}$ is dissipative and hence $m_B(\{Uq < \infty\}) = 0$. To show that $m_A(\{Uq = \infty\}) = 0$ it will suffice, by (1.9), to verify this whenever m_A is replaced by a potential μU with $\mu U \leq m_A$. But then we have $\infty > m(q) \geq \mu Uq$ and so $\mu(\{Uq = \infty\}) = 0$. Since $\{Uq < \infty\}$ is absorbing it follows that for each t, $\mu P_t(\{Uq = \infty\}) = 0$ and so $\mu U(\{Uq = \infty\}) = 0$.

The last thing we must do is record the fact that the $m^c + m^d$ decomposition is unique.

THEOREM 2. _If_ $m = m^c + m^d$ _where_ m^c _is conservative and_ m^d _is dissipative then_ $m^c = m_B$ _and_ $m^d = m_A$.

PROOF. Since B is absorbing, m_B^d is excessive. Suppose μU is a potential with $\mu U \leq m_B^d$. Then $\mu U \leq m_B$ and so μU is 0. Since m^d is the supremum of the potentials it dominates, we conclude that m_B^d is zero. Now $m_B = m_B^c + m_B^d = m_B^c$ and so by (1.7) m_B^c is invariant. Also $m^c = m_A^c + m_B^c$ and since m_B^c is invariant, m_A^c is excessive. Now $m_A^c \leq m_A$, and since m_A is the supremum of the potentials it dominates, if m_A^c is not zero there is a non zero potential μU with $\mu U \leq m_A^c \leq m^c$ contrary to the fact that m^c is conservative. So $m_A^c = m_B^d = 0$, which shows that $m^d = m_A$, and $m^c = m_B$.

References

1. Fitzsimmons, P. J. and Maisonneuve, B. "Excessive Measures and Markov Processes with Random Birth and Death", to appear in Z. Wahrscheinlichkeitstheorie verw. Gebiete.

2. Getoor, R. K. and Sharpe, M. J. "Naturality, standardness and weak duality for Markov processes." Z. Wahrscheinlichkeitstheorie verw. Gebiete, 67 (1984), 1-62.

3. Getoor, R. K. and Steffens, J. "Capacity Theory without Duality", to appear.

R. M. Blumenthal
Department of Mathematics
University of Washington
Seattle, Washington 98195

Seminar on Stochastic Processes, 1985
Birkhäuser, Boston, 1986

H^1 AND BMO SPACES OF ABSTRACT MARTINGALES

by

James K. Brooks and Nicolae Dinculeanu

1. Introduction

Abstract stochastic processes have been considered in various contexts by a number of authors. See, for example, Burkholder [2], Da Pratto [3], Kallianpur and Wolpert [14] and Métivier [15]. In this paper we shall examine the structure of H^1 and its dual BMO, for martingales taking their values in a Banach space E, and we use this to characterize weakly compact subsets in the former space. These results extend the theory for these Banach spaces developed by Dellacherie, Meyer, Yor and Mokobodzki [5]. The condition imposed on E is that it have the Radon-Nikodym property (RNP), which is not unexpected since this is a necessary and sufficient condition that the martingale convergence theorem holds in E. The connection between RNP and the geometry of E has been under intense study for over fifteen years in functional analysis. However, even without this assumption, by using the theory of lifting [13], a representation theorem for elements in $(H_E^1)'$ is proved (Theorem 3) (notation is given below). More precisely, every element of $(H_E^1)'$ is of the form $X \to E(\int \langle X_t, dA_t \rangle)$, where the optional process A with integrable variation has E' as its range. One of the tools used in the derivation is representation theory for operators on abstract Lebesgue spaces. This, together with the extended Herz-Lepingle theorem, establishes the desired duality theorem: $(H_E^1)' = BMO_{E'}$, and allows us to obtain criteria for weakly compact (conditional and relative) subsets $K \subset H_E^1$ by using the authors' earlier results on compactness [1], [8].

In § 2, the representation theorem of the dual of H_E^1 is proved; § 3 deals with $BMO_{E'}$ and the duality between H_E^1 and $BMO_{E'}$; compactness theorems are presented in § 4; the existence of a cadlag version of any E-valued martingale is proved in § 5, the Appendix, along with some facts about abstract martingales.

The notation and terminology are that used in Dellacherie and Meyer's Probabilities and Potential [4], and will be strictly adhered to. We shall present some notation here, but obvious modifications of scalar definitions will not be stated. Operator theory used in this paper uses the standard notation given in [12], [6].

$(\Omega, \mathscr{F}, (\mathscr{F}_t)_{t \geq 0}, P)$ is a stochastic base satisfying the usual conditions. $\mathscr{R}_E^1$ is the space of cadlag, separable valued processes, with values in E, having limits at $+\infty$ and satisfying $\|X\|_{\mathscr{R}_E^1} = \|X^*\|_{L^1} < \infty$. H_E^1 is the subspace of E-valued cadlag martingales X with norm $E(X^*) < \infty$. The definition of BMO_E is given in § 3. We remark that, as in [4], a stochastic process having integrable variation by definition means that it is cadlag. The variation $|A|_t$ of an E-valued process is defined by $|A|_t = \|A_0\| + |A|_{[0,t]}$, where $|A|_{[0,t]}$ is the variation of A on [0,t]; by convention, $A_{0-} \equiv 0$. For cadlag processes $X \in \mathscr{R}_E^1$ we define $X_\infty = X_{\infty -}$.

2. The space H_E^1 and a representation theorem for continuous linear functional on H_E^1.

The following two Propositions will be useful in examining the structure of H_E^1.

PROPOSITION 1. Let X be a cadlag, adapted, E-valued process such that $E(X^*) < \infty$. Then there exists a sequence of predictable stopping times T_n such that $\|X^{T_n-}\| \le n$ and $E(X - X^{T_n-})^* \to 0$.

As a consequence, the bounded martingales are dense in H_E^1.

PROOF. For each n, let $T_n = \inf\{t : X_{t-}^* > n\}$. Then each T_n is a predictable stopping time and $T_n \uparrow \infty$ a.s. Also $\|X^{T_n-}\| \le n$ and $\|X - X^{T_n-}\| \le 2X^*$. Since $\|X - X^{T_n-}\| \to 0$ a.s. and X^* is integrable, it follows that $(X - X^{T_n-})^* \to 0$ in L^1.

In particular, if $X \in H_E^1$ then $X^{T_n-} \in H_E^1$ (see appendix), and the last conclusion follows.

PROPOSITION 2. Suppose that $\mathscr{F}$ and E are separable. Then H_E^1 is separable.

PROOF. By Proposition 1, it suffices to consider only bounded martingales in H_E^1. If X is such a martingale, obtain a sequence (X_∞^n) of simple functions in L_E^∞ such that $X_\infty^n \to X_\infty$ a.s. and $\|X_\infty^n\| \le \|X_\infty\|$ for each n. Choose the simple functions X_∞^n over a countable field generating $\mathcal{F}$ with coefficients in a countable dense subset of E. Let $\mathcal{R}_\infty$ be the countable set of all such step functions, and let $\mathcal{R}$ be the corresponding set of cadlag martingales $Y_t = E(Y_\infty|\mathcal{F}_t)$, for $Y_\infty \in \mathcal{R}_\infty$. Note that $(X^n)^* \le \|X_\infty\|_\infty$. By Doob's inequality, $(X^n - X)^* \to 0$ in probability. We then apply the Vitali convergence theorem to conclude that $X^n \to X$ in H_E^1.

THEOREM 3. *Let* $J:H_E^1 \to R$ *be a continuous linear functional. There exists a stochastic function* $A:R_+\times\Omega \to E'$ *with raw integrable variation* $|A|$ *such that for every* $x \in E$, $\langle x,A\rangle$ *is optional, and the following representation holds:*

$$J(X) = E(\int\langle X_s,dA_s\rangle), \text{ for } X \in H_E^1.$$

Moreover $|A|$ *has its left potential bounded by* $2\|J\|$. *If, in addition,* E' *has the* RNP, *then* A *is optional.*

Conversely, if $A:R_+\times\Omega \to E'$ *is a stochastic function with raw integrable variation* $|A|$ *such that* $\langle x,A\rangle$ *is optional for every* $x \in E$ *and* $|A|$ *has its left potential bounded by a constant c, then for every* $X \in H_E^1$ *the integral* $E(\int\langle X_s,dA_s\rangle)$ *is defined and the mapping*

$$J(X) = E(\int\langle X_s,dA_s\rangle), \text{ for } X \in H_E^1$$

is a continuous linear functional on H_E^1 *with* $\|J\| \le c$. *If* $X \in H_E^1$ *is bounded, then*

$$J(X) = E(\langle X_\infty, A_\infty \rangle).$$

REMARK. If E is separable or if E' has the RNP, then $|A|$ is optional.

PROOF. We shall start with the proof of the second part of the theorem, which is shorter. Let A be a stochastic function satisfying the hypotheses of the second part of the theorem. Apply theorems VII.67 and VII.70 in [4] to the finite variation process $|A|$. Then for every $\Phi \in \mathscr{R}^1$, the integral $E(\int \Phi_s d|A|_s)$ is defined and the linear functional $J' : \mathscr{R}^1 \to R$ defined by

$$J'(\Phi) = E(\int \Phi_s d|A|_s), \text{ for } \Phi \in \mathscr{R}^1$$

is continuous and satisfies $\|J'\| \leq c$.

If $X \in \mathscr{R}_E^1$ then X is separably valued and $\|X\| \in \mathscr{R}^1$, hence $E(\|X_s\| d|A|_s) < \infty$. By theorem 5 in [10], the integral $E(\langle \int X_s, dA_s \rangle)$ is defined and the linear functional $J: \mathscr{R}_E^1 \to R$ defined by

$$J(X) = E(\int \langle X_s, dA_s \rangle)$$

is continuous:

$$|J(X)| \leq E(\int \|X_s\| d|A|_s) \leq c\|X\|_{\mathscr{R}^1}.$$

Hence $\|J\| \leq c$; in particular, the restriction of J to H_E^1 satisfies the same inequality. If $X \in H_E^1$ is bounded, then X is the optional projection of the constant process X_∞.

Since $\langle x,A\rangle$ is optional for every $x \in E$, we have, by theorem 8 in [11]

$$J(X) = E(\int\langle X_s, dA_s\rangle) = E(\int\langle X_\infty, dA_s\rangle)$$

$$= E(\langle X_\infty, \int_{[0,\infty]} dA_s\rangle) = E(\langle X_\infty, A_\infty\rangle).$$

To prove the first part of the theorem, let $J: H_E^1 \to R$ be a continous linear functional. In order to establish the existence of a stochastic function A satisfying the conditions of the theorem, we shall divide the proof into several steps.

1). By the Hahn-Banach theorem, we extend J to a continuous linear functional with the same norm on the space $\mathscr{R}_E^1$. We denote the extension by the same letter.

For every $\Phi \in \mathscr{R}_+^1$, we define $|J|(\Phi) = \sup\{|J(X)|: X \in \mathscr{R}_E^1$ $\|X\| < \Phi\}$. To see that $|J|$ is additive on $\mathscr{R}_+^1$, let $\Phi, \Psi \in \mathscr{R}_+^1$ and assume for the moment that $\Phi + \Psi > 0$ everywhere. Let $X \in \mathscr{R}_E^1$ with $\|X\| < \Phi + \Psi$. Then $Y = X\Phi(\Phi + \Psi)^{-1}$ and $Z = X\Psi(\Phi + \Psi)^{-1}$ belong to $\mathscr{R}_E^1$ and we have $\|Y\| < \Phi$, $\|Z\| < \Psi$, and $X = Y + Z$. It follows that $|J(X)| < |J|(\Phi) + |J|(\Psi)$, hence

$$|J|(\Phi + \Psi) < |J|(\Phi) + |J|(\Psi).$$

Now let Φ and Ψ be arbitrary in $\mathscr{R}_+^1$ and let $\varepsilon > 0$. By the above, it follows that

$$|J|(\Phi + \Psi) < |J|(\Phi) + |J|(\Psi) + \varepsilon\|J\|,$$

which establishes the desired inequality. The reverse inequality easily follows.

Since $|J|$ is positive and additive on $\mathscr{R}^1_+$, it can be extended to a continuous linear functional on $\mathscr{R}^1$, still denoted by $|J|$. We have

$$|J|(\Phi) \le \|J\|\|\Phi\|_{\mathscr{R}^1} \text{ for } \Phi \in \mathscr{R}^1_+$$

and

$$|J(X)| \le |J|(\|X\|), \text{ for } X \in \mathscr{R}^1_E.$$

Observe that J and $|J|$ have the same norm and $|J|$ is the smallest positive linear functional on $\mathscr{R}^1$ satisfying the above inequality.

2). We next use the device of forming two copies of Ω to represent J and $|J|$ as integrals (cf. [4]). More precisely, let Ω_- and Ω_+ be two disjoint copies of Ω and let $\tilde{\Omega} = \Omega_- \cup \Omega_+$. Let $W_- = (0,\infty]\times\Omega_-$, $W_+ = [0,\infty)\times\Omega_+$, $\tilde{W} = W_- \cup W_+$. For $X \in \mathscr{R}^1_E$ define $\tilde{X}:\tilde{W} \to E$ by $\tilde{X}(t,\omega_-) = X_{t-}(\omega)$, $\tilde{X}(t,\omega_+) = X_t(\omega)$; similar definitions for $\tilde{\Phi}$ when $\Phi \in \mathscr{R}^1$. Let $\tilde{\mathscr{R}}^1_E = \{\tilde{X}: X \in \mathscr{R}^1_E\}$ and $\tilde{\mathscr{R}}^1 = \{\tilde{\Phi}: \Phi \in \mathscr{R}^1\}$. Define $\tilde{J}: \tilde{\mathscr{R}}^1_E \to R$ by

$$\tilde{J}(\tilde{X}) = J(X) \text{ for } X \in \mathscr{R}^1_E.$$

Define also $|J|^\sim: \tilde{\mathscr{R}}^1 \to R$ by

$$|J|^\sim(\tilde{\Phi}) = |J|(\Phi), \text{ for } \Phi \in \mathscr{R}^1.$$

Define $|\mathfrak{J}|$ on $\tilde{\mathscr{R}}^1_+$ as in step 1), and observe that $|\mathfrak{J}| = |J|^{\sim}$.

The functional $|\mathfrak{J}|$ satisfies the Daniell condition on the set of bounded processes of $\tilde{\mathscr{R}}^1$, that is, if $\tilde{\Phi}^n \downarrow 0$, then $\Phi^n_- \downarrow 0$ and $\Phi^n \downarrow 0$; hence $\Phi^n \downarrow 0$ uniformly. Thus $\|\Phi^n\|_{\mathscr{R}^1} \to 0$ and consequently $|J|(\Phi^n) \to 0$, that is, $|\mathfrak{J}|(\tilde{\Phi}^n) \to 0$. By the Daniell theorem there exists a positive measure λ on $\tilde{W}$ endowed with the σ-algebra $\tilde{\mathscr{H}}$ generated by $\tilde{\mathscr{R}}^1$ such that

$$|\mathfrak{J}|(\tilde{\Phi}) = \int \tilde{\Phi} d\lambda, \text{ for } \Phi \in \tilde{\mathscr{R}}^1 \text{ bounded.}$$

By the method in [4], extend λ to the σ-algebra obtained by adjoining to $\tilde{\mathscr{H}}$ the evanescent sets of $\tilde{W}$, so that λ vanishes on evanscent sets - where evanescent is defined in the obvious way. Also extend λ to the σ-algebra obtained by adjoing to $\tilde{\mathscr{H}}$ the sets W_- and W_+ (use λ^* here).

We assert that $\tilde{\mathscr{R}}^1$ is a dense subset of $L^1(\lambda)$ and $|\mathfrak{J}|(\tilde{\Phi}) = \int \tilde{\Phi} d\lambda$, for $\Phi \in \mathscr{R}^1$. In fact, let $\tilde{\Phi} \in \tilde{\mathscr{R}}^1$; obtain a sequence of bounded processes in $\mathscr{R}^1$ such that $|\Phi^n| \le |\Phi|$ and $\Phi^n \to \Phi$ in $\mathscr{R}^1$. Extracting a subsequence, if necessary, we have $E((\Phi^n - \Phi)^*) \to 0$ and $(\Phi^n - \Phi)^* \to 0$ a.s. Hence, outside an evanescent set, (Φ^n) and (Φ^n_-) converge uniformly to Φ and Φ_- repspectively, which implies $\tilde{\Phi}^n - \tilde{\Phi} \to 0$ outside an evanescent set. It follows that $|J|(|\Phi^n - \Phi^m|) \to 0$ as $n,m \to \infty$; hence $\int |\tilde{\Phi}^n - \tilde{\Phi}^m|)d\lambda \to 0$, that is $(\tilde{\Phi}^n)$ is Cauchy in $L^1(\lambda)$. Thus $\tilde{\Phi} \in L^1(\lambda)$; consequently, $|\mathfrak{J}|(\tilde{\Phi}) = \int \tilde{\Phi} d\lambda$. By examining the construction of the Daniell integral, the set of bounded functions of $\tilde{\mathscr{R}}^1$ can be shown to be dense in $L^1(\lambda)$, which completes this step.

3). $\tilde{\mathscr{R}}^1_E$ is dense in $L^1_E(\lambda)$. In fact, each $\tilde{X} \in \tilde{\mathscr{R}}^1_E(\lambda)$ is separably valued and $\|\tilde{X}\| \in \tilde{\mathscr{R}}^1 \subset L^1(\lambda)$, hence $\tilde{X} \in L^1_E(\lambda)$; therefore $\tilde{\mathscr{R}}^1_E \subset L^1_E(\lambda)$. Let $\tilde{\Phi} \in L^1(\lambda)$; by step 2), obtain a sequence of bounded $\tilde{\Phi}^n \in \tilde{\mathscr{R}}^1$ such that for $x \in E$ we have $\tilde{\Phi}^n x \to \tilde{\Phi} x$ in $L^1_E(\lambda)$. Since $\{\Sigma\tilde{\Phi}^i x_i : x_i \in E, \tilde{\Phi}^i \in L^1(\lambda)\}$ is dense in $L^1_E(\lambda)$, and these functions can be approximated by bounded functions of $\tilde{\mathscr{R}}^1_E$ the assertion follows.

We also have $|\tilde{J}(\tilde{X})| \leq \int \|\tilde{X}\| d\lambda$, for $\tilde{X} \in \tilde{\mathscr{R}}^1_E$; therefore we can extend $\tilde{J}$ to $L^1_E(\lambda)$, still denoted by $\tilde{J}$, which satisfies the above inequality for all $\tilde{X} \in L^1_E(\lambda)$. Then there exists a measure $m: \tilde{\mathscr{H}} \to E'$ with finite variation, $|m| \leq \lambda$ (Cor. 2, p.261, in [6]) satisfying

$$\tilde{J}(\tilde{X}) = \int \tilde{X} dm \text{ for } \tilde{X} \in L^1_E(\lambda).$$

Moreover $|m| = \lambda$. To see this, let $J'(\Phi) = \int \tilde{\Phi} d|m|$, for $\Phi \in \mathscr{R}^1$. Note that J' is a positive linear functional on $\mathscr{R}^1$ dominating J; hence $|J| \leq J'$. On the other hand, observe that $J'(\Phi) \leq |J|(\Phi)$, hence $J' = |J|$. As a result $|m| = \lambda$ and $|\tilde{J}|(\tilde{\Phi}) = \int \tilde{\Phi} d|m|$, for $\tilde{\Phi} \in L^1(\lambda)$.

4). We decompose $\lambda = \lambda_- + \lambda_+$, $m = m_- + m_+$, where λ_-, m_- are equal to λ, m on W_- endowed with the predictable σ-algebra $\tilde{\mathscr{H}} \cap W_-$, and 0 outside W_- and λ_+, m_+ are equal to λ, m on W_+ endowed with the optional σ-algebra $\mathscr{H} \cap W_+$ and 0 outside W_+. The variations of m_- and m_+ on $\mathscr{H}$ are respectively λ_- and λ_+.

Since W_- and W_+ are disjoint, we have $|m| = \lambda_- + \lambda_+$. Thus

$$J(X) = \int_{W_-} X_- dm_- + \int_{W_+} X dm_+ \text{ for } X \in \mathscr{R}^1_E$$

and

$$|J|(\Phi) = \int_{W_-} \Phi_- d\lambda_- + \int_{W_+} \Phi d\lambda_+, \text{ for } \Phi \in \mathscr{R}^1.$$

5). Identify the measurable space $(W_-, \tilde{\mathscr{H}} \cap W_-)$ with $((0,\infty]\times\Omega, \mathscr{P})$, and consider m_- and λ_- as being defined on $\mathscr{P}$; then consider their predictable extensions to $\mathscr{B}((0,+\infty])\times\mathscr{F}$, (denoted by the same letters), and finally to $\mathscr{B}([0,\infty]\times\mathscr{F}$ by setting λ_- and m_- to be zero on $\{0\}\times\Omega$. Similarly, we consider m_+ and λ_+ defined on $\mathscr{O}$, then we consider optional extensions to $\mathscr{B}([0,\infty))\times\mathscr{F}$ (denoted by the same letters) and finally to $\mathscr{B}([0,\infty])\times\mathscr{F}$ by setting m_+ and λ_+ equal to zero on $\{\infty\}\times\Omega$. Denote by $|m_+|$ and $|m_-|$ respectively the variations of m_+ and m_- on $\mathscr{B}([0,\infty])\times\mathscr{F}$, by $|m_+|^o$ the optional projection of $|m_+|$ and by $|m_-|^p$ the predictable projection of $|m_-|$. Then $|m_+|^o = \lambda_+$ and $|m_-|^p = \lambda_-$. In fact, let $B \in \mathscr{O}$ and let (B_i) be a finite family of disjoint sets from $\mathscr{O}$ contained in B. Then $\Sigma\|m_+(B_i)\| \leq \Sigma|m_+|(B_i) \leq |m_+|(B)$; therefore $\lambda_+(B) \leq |m_+|(B)$. If now, (C_i) is a finite family of disjoint sets from $\mathscr{B}([0,\infty])\times\mathscr{F}$ contained in B we have:

$$\sum\|m_+(C_i)\| = \Sigma\|m_+({}^o1_{C_i})\| \leq \sum \lambda_+({}^o1_{C_i}) = \sum \lambda_+(C_i) \leq \lambda_+(B);$$

therefore $|m_+|(B) \leq \lambda_+(B)$. It follows that $|m_+| = \lambda_+$ on $\mathscr{O}$, hence $|m_+|^o = \lambda_+$; and we can prove similarly that $|m_-|^p = \lambda_-$. Since for $\Phi \in \mathscr{R}^1$, Φ is optional and Φ_- is predictable, we have

$$|J|(\Phi) = \int\Phi_- d\lambda_- + \int\Phi d\lambda_+ = \int\Phi_- d|m_-| + \int\Phi d|m_+|.$$

6). By theorem 6 in [10], there is a stochastic function $A^-: [0,\infty]\times\Omega \to E'$, with $A_0^- = 0$, which has raw integrable variation $|A^-|$ such that $\langle x,A^-\rangle$ is predictable for $x \in E$ and satisfies

$$m_-(Y) = E(\int_{[0,\infty]}\langle Y_s, dA_s^-\rangle), \text{ for } Y \in L_E^1(|m_-|)$$

and

$$|m_-|(\Phi) = E(\int_{[0,\infty]}\Phi_s d|A^-|_s), \text{ for } \Phi \in L_E^1(|m_-|).$$

Also there exists a stochastic function $A^+: [0,\infty]\times\Omega \to E'$ with $A_{\infty-}^+ = A_\infty^+$ with raw integrable variation $|A^+|$ such that $\langle x,A^+\rangle$ is optional for $x \in E$ and satisfies

$$m_+(Z) = E(\int_{[0,\infty]}\langle Z_s, dA_s^+\rangle), \text{ for } Z \in L_E^1(|m_+|)$$

and

$$|m_+|(\Phi) = E(\int_{[0,\infty]}\Phi_s d|A^+|_s), \text{ for } \Phi \in L^1(|m_+|).$$

It follows that

$$J(X) = E(\int_{[0,\infty]}\langle X_{s-}, dA_s^-\rangle + \int_{[0,\infty]}\langle X_s, dA_s^+\rangle), \text{ for } X \in \mathscr{R}_E^1$$

and

$$|J|(\Phi) = E(\int_{[0,\infty]}\Phi_{s-} d|A^-|_s + \int_{[0,\infty]}\Phi_s d|A^+|_s), \text{ for } \Phi \in \mathscr{R}^1.$$

Note that, if E' has the RNP, then A^- is predictable, A^+ is optional, and these functions are the unique stochastic functions satisfying the above equalities. In this case $|A^-|$ is predictable and $|A^+|$ is optional by right continuity of A^- and A^+. In the case E' does not have the RNP, if ρ is a lifting of P, then A^- and A^+ can be chosen uniquely up to an evanescent set such that $\rho[A_t^-] = A_t^-$ and $\rho[A_t^+] = A_t$ for every t (see [6] p. 212).

We note also that by theorem VII.67 in [4], applied to $|J|$, $|A^-|$ and $|A^+|$, for every stopping time T, we have

$$E(\int_{(T,\infty]} d|A^-|_s + \int_{[T,\infty]} d|A^+|_s | \mathscr{F}_t) \leq \| |J| \| = \|J\| \text{ a.s.}$$

7). Let $A = A^- + A^-$. For every martingale $X \in H_E^1$, X_- is the predictable projection of X. Since $\langle x, A^- \rangle$ is predictable for each x, we have

$$\int_{[0,\infty]} \langle X_{s-}, dA_s^- \rangle = \int_{[0,\infty]} \langle X_s, dA_s^- \rangle,$$

hence

$$J(X) = E(\int_{[0,\infty]} \langle X_s, dA_s \rangle).$$

Since $\rho[A_t^-] = A_t^-$ and $\rho[A_E^+] = A_t^+$, we see that $\rho[A_t] = A_t$, see [6, p.212]. By proposition 5, page 213 in [6], it follows that $\|A_t\|$ is $\mathscr{F}$-measurable. By the right continuity of A, using partitions consisting of rational points, we deduce that $|A|_t$ is $\mathscr{F}$-measurable for each t. Since $|A|$ is right continuous, $|A|$ is $\mathscr{B}[0,\infty] \times \mathscr{F}$-measurable, and $|A| \leq |A^-| + |A^+|$ implies that $|A|$ is integrable (but not necessarily adapted).

Let $Z(B)$ and $Z^g(B)$ respectively denote the potential and left potential generated by an increasing process B.

For any stopping time T we have

$$Z_T(|A|) \leq Z_T([|A^-| + |A^+|]) \leq \|J\| \text{ a.s.}$$

by step 6). Then by VII.68 in [4],

$$Z_T^g(|A|) \leq 2\|J\| \text{ a.s.}$$

8). If E is separable, or if E' has the RNP then $|A|$ is optional. In fact for each $t \geq 0$ we have $|A|_t = \|A_0\| + \sup_n \sum_{1 \leq k < \infty} \|A_{(k+1)2^{-n}\wedge t} - A_{k2^{-n}\wedge t}\|$ by right continuity of A. If E is separable, then for each $a \geq 0$ and $b \geq 0$, the stochastic function $(\|A_{a\wedge t} - A_{b\wedge t}\|)_{t\geq 0}$ is optional; the same is true if E' has the RNP, since in this case A is optional. It follows that in both cases $|A|$ is optional.

3. The space BMO_E.

In this section we introduce the space BMO_E and prove that the dual of H^1_E is $BMO_{E'}$.

DEFINITION. We denote by BMO_E the vector space of E-valued cadlag martingales $Y_t = E(Y_\infty|\mathscr{F}_t)$ satisfying the following condition:

There exists a constant c such that $E(\|Y_\infty - Y_{T-}\| |\mathscr{F}_T) \leq c$ a.s. for every stopping time T. The smallest constant c is denoted by $\|Y\|_{BMO}$ and this defines a semi norm on BMO_E.

The following proposition, an extension of the Herz-Lepingle theorem, yields a characterization of BMO_E in terms of processes with integrable variation, which will be used later in the characterization of the dual of H^1_E as $BMO_{E'}$, when E' has the RNP. The proof of this proposition follows along the lines of VII.70, VII.76, VII.77 and VII.78 in [4], using results listed in the Appendix. We shall omit the lengthy technical details.

PROPOSITION 4. Let $Y_t = E(Y_\infty|\mathcal{F}_t)$ be an E-valued cadlag martingale. The following three assertions are equivalent:

(a) $Y \in BMO_E$;

(b) There exists an E-valued process with integrable variation $|A|$ such that $A_\infty = Y_\infty$ a.s. and $|A|$ generates a bounded left potential;

(c) There exists an E-valued raw process B with bounded variation, possesing optional dual projection B^o satisfying $(B^o)_\infty = Y_\infty$ a.s.

Moreover if $c > 1$ is given, and if $\|\Delta Y\| \leq a$, and $E(\|Y_\infty - Y_T\| |\mathcal{F}_T) \leq b$ for every stopping time T, then we can choose A such that $Z^g(|A|) \leq b(1 + \frac{c+a}{c-1})$. On the other hand, if $Z^g(|A|) \leq c$, then $\|\Delta Y\| \leq 2c$, $E(\|Y_\infty - Y_t\| |\mathcal{F}_t) \leq 2c$ and $E(\|Y_\infty - Y_{T-}\| |\mathcal{F}_T) \leq 3c$.

THEOREM 5. Let E, F and G be Banach spaces and let $(u,v) \to uv$ be a bilinear continuous mapping of $E\times F$ into G.

If $Y \in BMO_E$ *then there exists a constant* k *such that*

$$\|E(X_\infty Y_\infty)\| \leq kE(X^*)$$

for every bounded martingale $X \in H^1_F$, *where* $k = 2b(3 + a)$.

PROOF. Let A be the process in (b) of the preceeding proposition. Use the fact that A and X^* are optional and "integration by parts" in [4] to deduce

$$\|E(X_\infty Y_\infty)\| = \|E\int_0^\infty X_s dA_s\| \leq$$

$$\leq E(\int_0^\infty X_s^* d|A|_s) = E(\int_0^\infty (|A|_\infty - |A|_{s-})dX_s^*) =$$

$$= E(\int_0^\infty {}^o(|A|_\infty - |A|_-)_s dX_s^*) =$$

$$E(\int_0^\infty E(|A|_\infty - |A|_{s-}|\mathcal{F}_s)dX_s^*) \leq kE(X^*).$$

$k = 2b(3 + a)$ follows by choosing $c = 2$.

THEOREM 6. *Assume* E' *has the* RNP. *Let* Y *be an* E'*-valued martingale. If there is a constant* k *such that*

$$|E(\langle X_\infty, Y_\infty\rangle)| \leq kE(X^*)$$

for every bounded martingale $X \in H^1_E$ *then* $Y \in BMO_E$, *and* $\|Y\|_{BMO} \leq 2k$.

PROOF. $J(X) = E(\langle X_\infty, Y_\infty\rangle)$ defines a linear continuous functional on a dense (Proposition 1) subset of H^1_E. Denote its extension to H^1_E also by J. By Theorem 3 there is an E'-valued process A with integrable variation $|A|$ satisfying

$$J(X) = E(\int_0^\infty \langle X_s, dA_s\rangle), \text{ for } X \in H^1_E$$

and

$$Z^g(|A|) < 2k.$$

Moreover, for $X \in H^1_E$ bounded, we have (see [11], theorem 9),

$$E(\langle X_\infty, Y_\infty\rangle) = E(\int_0^\infty \langle X_s, dA_s\rangle) =$$

$$= E(\int_0^\infty \langle {}^O(X_\infty), dA_s\rangle) = E(\langle X_\infty, A_\infty\rangle).$$

Since X_∞ is arbitrary and bounded in L^∞_E we deduce that $Y_\infty = A_\infty$ a.s. The conclusion follows from Proposition 4.

The following theorem follows from the preceeding results; the operator norm $|||\cdot|||$ is defined in [6]. The constant 10 in the first inequality is an upper bound of the constants in Proposition 4.

THEOREM 7. _Let_ E, F, G _be as in theorem_ 5. _For every_ $Y \in BMO_E$ _there is a continuous operator_ $J: H^1_F \to G$ _with_ $J(X) = E(X_\infty Y_\infty)$, _for_ $X \in H^1_F$ _bounded and_
$|||\, J\, ||| < 10\|Y\|_{BMO}$.

Conversely if E' _has the_ RNP, _for every continuous linear functional_ $J: H^1_E \to R$, _there is a_ $Y \in BMO_{E'}$ _such that_

$$J(X) = E(\langle X_\infty Y_\infty \rangle) \text{ for } X \in H^1_E \text{ bounded}$$

and

$$\|Y\|_{BMO} \le 2\|J\|.$$

COROLLARY. *If* E' *has the* RNP *then the dual of* H^1_E *is norm equivalent to* $BMO_{E'}$.

4. **Weak Compactness in H^1_E.**

In this section we shall always assume that E' has the RNP. General results concerning weak compactness in L^1_E can be found in the authors' paper [1]; conditional weak compactness means that every sequence has a weak Cauchy subsequence. The following theorem extends results in [5].

THEOREM 8. *A set* $K \subset H^1_E$ *is conditionally weakly compact if and only if:*

1) $K^* = \{X^*: X \in K\}$ *is uniformly integrable;*

2) *For each* $A \in \mathscr{F}$, $K_\infty(A) = \{\int_A X_\infty dP: X \in K\}$ *is conditionally weakly compact in* E.

PROOF. Assume first that K is conditionally weakly compact. Since the mapping $X \to \int_A X_\infty dP$ of H^1_E into E is a continuous operator, it is continuous for the weak topologies, and 2) follows.

To prove 1), note that for every random variable $S > 0$, the operator $X \to X_S$ on H^1_E into L^1_E is continuous, hence $K_S = \{X_S: X \in K\}$ is conditionally weakly compact in L^1_E;

hence K_S is uniformly integrable by theorem 1 in [1]. Since K is bounded in H_E^1, lemma 5 in [3] implies 1).

Conversely, assume conditions 1) and 2). Then K_∞ is uniformly integrable in L_E^1; it follows by 2) that K_∞ is conditionally weakly compact in L_E^1 (see [1]).

Choose a sequence (X^n) from K. Then (X_∞^n) contains a weak Cauchy subsequence in K_∞; assume that the full sequence has this property. Note that $(X^n)^*$ is uniformly integrable. To show that (X^n) is weak Cauchy in H_E^1, we need the following lemma.

LEMMA 9. <u>If</u> $(X^i)_{i\in I}$ <u>is a net in</u> H_E^1 <u>such that</u> $((X^i)^*)_{i\in I}$ <u>is uniformly integrable and</u> $(X_\infty^i)_{i\in I}$ <u>is a weak Cauchy net in</u> L_E^1 <u>then</u> $(X^i)_{i\in I}$ <u>is a weak Cauchy net in</u> H_E^1. <u>If</u> $X_\infty^i \to X_\infty$ <u>weakly in</u> L_E^1 <u>then</u> $X^i \to X$ <u>weakly in</u> H_E^1.

PROOF. Let J be a continuous linear functional on H_E^1 and let A be an optional E'-valued process associated with J by theorem 3. For each n, define a sequence of stopping times $T_n \uparrow \infty$ by $T_n = \inf\{t\colon |A|_t > n\}$. Set $A^n = A^{T_n-}$. The optional process A^n has finite variation $|A^n| \le n$. Moreover since for each stopping time T,

$$E(|A_\infty^n - A_{T-}^n|\,|\mathscr{F}_T) \le 2n,$$

the functional J^n associated with A^n is continuous on H_E^1. Note that $J^n(X^i) = E(\langle X_\infty^i, A_\infty^n\rangle)$. For fixed n, $(J^n(X^i))_{i\in I}$ is Cauchy. Also

$$|J(X^i) - J^n(X^i)| = |E(\int_{[T_n,\infty]}\langle X_s^i, dB_s\rangle)| \le$$

$$\leq E((X^i)^*|B|_\infty 1_{\{T_n<\infty\}}) \underset{n}{\to} 0$$

uniformly in i. To see this, recall that $(X^i)^*_{i\in I}$ are uniformly integrable; B is a bounded raw process with integrable variation such that $A = B^o$. Since $\{T_n < \infty\} \downarrow \phi$, the uniform convergence follows. Consequently $(J(X^i))_{i\in I}$ is Cauchy. The last statement in the conclusion now follows. This completes the proof of the lemma and of the theorem.

THEOREM 10. *If* E *is weakly sequentially complete, then* H^1_E *is weakly sequentially complete*.

PROOF. Let (X^n) be weak Cauchy in H^1_E. Then (X^n) is bounded, $((X^n)^*)$ is uniformly integrable and (X^n_∞) is weak Cauchy in L^1_E. By Tallagrand's theorem [17], $X^n_\infty \to X_\infty$ weakly for some $X_\infty \in L^1_E$. Let X be a cadlag version of the martingale $E(X_\infty|\mathscr{F}_t)$. If $E(X^*) < \infty$, we have $X \in H^1_E$ by the above lemma. However, this follows from the fact that the mapping $Z \to E(Z^*)$ is lower semi continuous on L^1_E for the weak topology. The proof for this is similar to that of lemma 2 in [5], using the equality

$$E(\int_{[0,\infty]}\|Z_s\|dA_s) = \sup E(\int_0^\infty \langle Z_s, Z'_s\rangle dA_s)$$

for $\|Z'\| \leq 1$, $Z' \in L^\infty_{E'}(\mu_A, \mathscr{O})$, where A is any increasing integrable process and μ_A is the P-measure generated by A, $\mu_A(M) = E(\int_{[0,\infty]} 1_M dA_s)$, for $M \in \mathscr{B}(R_+) \times \mathscr{F}$. This completes the proof.

THEOREM 11. *A sequence* (X^n) *of elements from* H_E^1 *converges weakly to zero if and only if:*

1) (X^n) *is bounded in* H_E^1 ;

2) $X_S^n \to 0$ *weakly in* L_E^1 *for every random variable* $S > 0$.

PROOF. If $X^n \to 0$ weakly in H_E^1 then (X^n) is bounded in H_E^1 and condition 2) follows from the continuity of the mapping $X \to X_S$ of H_E^1 into L_E^1.

Conversely, assume 1) and 2). Then for each $S > 0$, (X_S^n) is uniformly integrable, which implies, by lemma 5 in [5] that $((X^n)^*)$ is uniformly integrable. Take $S \equiv \infty$ and apply Lemma 9 to obtain the conclusion.

COROLLARY 12. $X^n \to 0$ *weakly in* H^1 *if and only if:*

1) $\sup_n E((X^n)^*) < \infty$

2) $E(1_A X_S^n) \to 0$ *for every r.v.* $S > 0$ *and* $A \in \mathcal{F}$.

THEOREM 13. (X^n) *is weak Cauchy in* H_E^1 *if and only if:*

1) (X^n) *is bounded in* H_E^1 ;

2) (X_S^n) *is weak Cauchy in* L_E^1 *for every r.v.* $S > 0$.

We close this section with criteria for weak compactness in terms of weak uniform convergence of certain nets of operations on H_E^1. If $\mathcal{G}$ is a sub σ-algebra of $\mathcal{F}$ and X is an element of H_E^1 then $E_{\mathcal{G}}X$ or $E(X|\mathcal{G})$ will denote the cadlag version of the martingale $E(X_t|\mathcal{G})_{t \in R_+}$. Note that $E_{\mathcal{G}}$ is a contraction of H_E^1 into itself. First of all we need the following result.

PROPOSITION 14. Let $(\mathscr{G}_\alpha)$ be an increasing net of sub σ-algebras of $\mathscr{F}$, and $\mathscr{G} = \bigvee \mathscr{G}_\alpha$. If $X \in H^1_E$ then $\lim_\alpha E_{\mathscr{G}_\alpha} X = E_{\mathscr{G}} X$ strongly in H^1_E.

PROOF. If X is bounded in H^1_E then $E(X_\infty|\mathscr{G}_\alpha) \underset{\alpha}{\to} E(X_\infty|\mathscr{G})$ in L^1_E. Use Doob's inequality and the dominated convergence theorem for nets (convergence in probability is used here because of net convergence) to conclude that the proposition is true for X bounded. The general result follows using Proposition 1 and the Banach-Steinhauss theorem.

THEOREM 15. Let $(\mathscr{G}_n)$ be an increasing sequence of sub σ-algebras generating $\mathscr{F}$. If the set $K \subset H^1_E$ is relatively weakly compact then:

1) Each $E_{\mathscr{G}_n} K$ is relatively weakly compact;

2) $\lim_n E_{\mathscr{G}_n} X = X$ weakly in H^1_E uniformly for $X \in K$.

Conversely, even if E' does not have the RNP, if $K \subset H^1_E$ satisfies 1) and 2), then K is conditionally weakly compact.

PROOF. Assume that K is relatively weakly compact. Condition 1) follows from the continuity of each $E_{\mathscr{G}_n}$. To prove 2), let K_b the set of all bounded elements $Y \in H^1_E$ such that $Y^* \leq X^*$ for some $X \in K$. By Theorem 8, it follows that K_b^* is uniformly integrable; we shall prove 2) for $X \in K_b$. Let J belong to the dual of H^1_E and let A be a stochastic function representing J. Recall that for $X \in K_b$ we have $J(X) = E(\langle X_\infty, A_\infty\rangle)$ and that the martingales $E_{\mathscr{G}_\alpha} X \equiv E_\alpha X$ are bounded.

For $X \in K_b$ we have

$$|J(E_\alpha X - X)| = |E(\langle (E_\alpha X)_\infty - X_\infty, A_\infty \rangle)| =$$

$$= |E(\langle X_\infty, E_\alpha(A_\infty)\rangle) - E(\langle X_\infty, A_\infty\rangle)| <$$

$$< E(X^* Z E_\alpha(|A|_\infty)) + E(X^* Z |A|_\infty) + \lambda E(|E_\alpha(A_\infty) - A_\infty|),$$

where $Z = 1_{\{X^* > \lambda\}}$. By the uniform integrability of K_b the first two terms above can be made small when λ is large, uniformly on K_b. Then choose α_0 large enough so that the last term is also small for $\alpha > \alpha_0$. Since K_b is dense in K, the result follows.

Conversely, let K satisfy 1) and 2). Let $S > 0$ be a random variable on Ω. One can see that the set $(E_{\mathscr{G}_n} K)_S = \{(E_{\mathscr{G}_n} X)_S: X \in K\}$ is relatively weakly complact in L^1_E. Then by [1], [8], $\lim (E_{\mathscr{G}_n} X)_S = X_S$ weakly in L^1_E uniformly for $X \in K$. This implies K_S is conditionally weakly compact by using lemma 6 in [1].

5. Appendix

Here we show the existence of cadlag modifications of an E-valued martingale. Theorems 17 and 18 can be found in [16]. We give a new proof of Theorem 17 which seems to be easier than the existing proofs.

THEOREM 16. <u>Any martingale with values in a Banach space</u> E <u>has a cadlag modification</u>.

PROOF. Let (X_t) be an E-valued martingale. We have $\lim_{r \downarrow t} X_r = X_t$ in L^1_E (see theorem 17 and the remarks below). Since each X_r is Bochner integrable, there exists a separable subspace E_0 of E such that $X_r(\omega) \in E_0$ a.s. for all rational r. Hence $X_t(\omega) \in E_0$ a.s. for all $t > 0$. By using a modification of (X_t) we may therefore assume that the range of (X_t) is a separable Banach space E. Define $\Psi: E \to R^N$ by $\Psi(x) = (\|x - a_n\|)_{n\in N}$ where (a_n) is dense in E; Ψ is a homeomorphism of E onto the space $\Psi(E)$ endowed with the product topology. For each n, $\|X_t - a_n\|$ is a submartingale, and since $X_r \to X_t$ in L^1_E as $r \downarrow t$, we see that $t \to E(\|X_t - a_n\|)$ is right continuous. Hence the process $\|X_t - a_n\|$ has a cadlag modification (Y^n_t). Let $Y_t(\omega) = (Y^n_t(\omega))_n$ be the R^N-valued cadlag process satisfying $Y_t(\omega) = \Psi(X_t(\omega))$ a.s.. Let N_0 be a negligible subset of Ω such that $Y_r(\omega) = \Psi(X_r(\omega))$ for $\omega \notin N_0$ and rational r. For $\omega \notin N_0$, the right continuity of (Y_t) ensures that $Y_t(\omega) \in \Psi(E)$ for $\omega \notin N_0$ for all $t > 0$; redefine Y to be $(\|a_n\|)_n$ on $R_+ \times N_0$. Then $(\Psi^{-1}(Y_t))$ is a cadlag modification of (X_t).

THEOREM 17. (a) Let $\mathcal{B}_n$ be an increasing sequence of σ-algebras generating $\mathcal{B}_\infty$. Let X be an E-valued $\mathcal{B}_\infty$-integrable function, and $X_n = E_{\mathcal{B}_n}(X)$, $1 \le n \le \infty$. Then $X_n \to X_\infty$ a.s. and in L^1_E.

(b) Let $(X_n)_{n\le 0}$ be an E-valued martingale relative to the σ-algebras $\mathcal{B}_n$, and let $\mathcal{B}_{-\infty} = \cap \mathcal{B}_n$. Then $X_n \to E_{\mathcal{B}_{-\infty}}(X_0)$ a.s. and in L^1_E.

PROOF. (a) Define the operators $T_n = E_{\mathcal{B}_n}$ on L^1_E which are pointwise bounded and $T_n(X) = X$ on the dense subset $\bigcup_n L^1_E(\mathcal{B}_n)$ of $L^1_E(\mathcal{B}_\infty)$. By the uniform boundedness theorem, $T_n(X) \to X$ in $L^1_E(\mathcal{B}_\infty)$. By Doob's inequality, for fixed n_0,

$$P(\sup_{n>n_0} \|X_n - X_{n_0}\| > c) < \frac{1}{c} \sup_{n>n_0} E(\|X_n - X_{n_0}\|),$$

and for fixed c, the L^1_E convergence of X_n makes the right hand side small for large n_0. This shows that $X_n \to X_\infty$ a.s.

(b) Let $\Delta = \{X \in L^1_E(\mathcal{B}_0)$ satisfying the conclusion of the theorem$\}$. Again, let $T_n(X) = E_{\mathcal{B}_n}(X) = X_n$ for $-\infty < n < 0$. For $x \in E$, $A \in \mathcal{B}_0$, we have $x1_A \in \Delta$ by the classical convergence theorem for scalar martingales. Since (T_n) is pointwise bounded and $T_n(X) \to X_{-\infty}$ on a dense subset of $L^1_E(\mathcal{B}_0)$, we have $X_n \to X_{-\infty}$ in $L^1_E(\mathcal{B}_0)$, using the uniform boundedness theorem. By Doob's inequality again, it follows that $X_n \to X_{-\infty}$ a.e.

THEOREM 18. A Banach space E has the RNP if and only if every L^1_E bounded martingale converges a.s. to an integrable random variable.

REMARKS. In view of theorem 17, the classical theorems for continuous parameter martingales can be proved using convergence theorems in L^1_E (see [12]). For example, if X is an E-valued martingale and T is a (predictable) stopping time, then $X^T(X^{T-})$ is a martingale. All of the results in chapter VI in [4] go over in this more general setting.

References

1. J.K. Brooks and N. Dinculeanu, Weak compactness in the space of Bochner integrable functions and applications, Advances in Math. 24 (1977), 172-188.

2. D. Burkholder, A geometrical characterization of Banach spaces in which martingale difference sequences are unconditional, Annals of Prob. 9 (1981), 997-1011.

3. G. da Prato, Some results on linear stochastic differential equations in Hilbert spaces, Stochastic 6 (1982), 315-322.

4. C. Dellacherie and P.A. Meyer, Probabilities and Potential, North Holland, 1978, 1982.

5. C. Dellacherie, P.A. Meyer and M. Yor, Sur certaines propriétés des espaces de Banach H^1 et BMO, Séminaire de Probabilités, XII (1976-77), Springer, Lecture Notes, 649, 98-113.

6. N. Dinculeanu, Vector measures, Pergamon Press, Oxford, 1967.

7. N. Dinculeanu, Weak compactness and uniform convergence of operators in spaces of Bochner integrable functions, Journal of Math. Analysis and Applications, (1985).

8. N. Dinculeanu, Characterization of weak compactness in function spaces by means of uniform convergence of extended operators, Proc. of the Conference on Banach spaces, Columbia, Missouri, 1984, Springer Lecture Notes, (1985).

9. N. Dinculeanu, Weak compactness in the space H^1 of martingales, Séminaire de Probabilités XIX, Springer Lecture Notes 1123, 1985.

10. N. Dinculeanu, Vector valued stochastic processes I. Vector measures and vector valued processes with finite variation. (Preprint).

11. N. Dinculeanu, Vector valued stochastic processes III. Projections and dual projections. (Preprint).

12. N. Dunford and J. Schwartz, Linear operators Part I, Interscience, 1957.

13. A. and C. Ionescu Tulcea, Topics in the theory of lifting, Springer, 1969.

14. G. Kallianpur, and R. Wolpert, Infinite dimensional stochastic differential equation models for spatially distributed neurons (Preprint).

15. M. Métivier, Semimartingales, Walter de Gruyter, Berlin, 1982.

16. J. Neveu, Discrete-Parameter Martingales, North-Holland, Amsterdam, 1975.

17. M. Talagrand, Weak Cauchy Sequences in $L^1(E)$, Amer. J. Math. 106(1984), 703-724.

J. K. Brooks and N. Dinculeanu
Department of Mathematics
University of Florida
Gainesville, Florida 32611

Seminar on Stochastic Processes, 1985
Birkhäuser, Boston, 1986

BROWNIAN EXCURSIONS AND MINIMAL THINNESS

PART II

APPLICATIONS TO BOUNDARY BEHAVIOR OF THE GREEN FUNCTION

by

Krzysztof Burdzy

Introduction

The article presents what seems to be a new approach to studying the boundary behavior of the Green functions in subdomains of R^n, $n > 2$. This involves applications of probability, in particular Brownian excursion laws. The new method seems to be especially fruitful in application to the angular derivative problem (see Burdzy (1985b)).

The boundary behavior of the Green and related functions has been studied extensively and it is not prudent to claim originality. It can be said however that the results of this article are stronger than the corresponding results of Widman (1967). It has been pointed out to the author that some theorems of Ziemer (1974) and Jerison and Kenig (1982, 1983) may also yield similar estimates of the Green function.

The main results of this paper are Corollaries 4.1-4.4.

The present article is the middle part of a three part sequence (see Burdzy (1985a, b)). The reader is referred to Part I for definitions, notation and some lemmas. For the ease of reference there is a continuous numbering of sections, formulae and theorems throughout all three articles of the sequence.

The author would like to thank Carlos Kenig for the most useful advice.

4. Boundary behavior of the Green function

If D is a halfspace then the Green function $G_D(x,\cdot)$ has a nondegenerate derivative at every boundary point. How smooth should the boundary ∂D be near a point $y \in \partial D$ so that $G_D(x,\cdot)$ behaves like a linear function in a neighborhood of y? A precise statement of the problem and a partial answer are given below.

The "constants" c, c_1, c_2, ... will be numbers which may depend only on the dimension n of the space and the Lipschitz constants of functions considered below. They may take different values in different theorems, lemmas and proofs. The following notation will be used throughout the paper: $\mathbf{e} = (1,0,\ldots,0)$, $D_1 = \{x \in R^n : x_1 > 0\}$, $L = \partial D_1$. Unless indicated otherwise the integrals are taken with respect to surface area measure (denoted dx) on hyperplanes.

Let $D \subset R^n$ be a region such that $0 \in \partial D$. Assume that

(4.1) $\{x \in R^n : |x| < 1/c$ and

$$x_1 > c\sqrt{x_2^2 + x_3^2 + \ldots + x_n^2}\} \subset D$$

for some c, $0 < c < \infty$. Fix an arbitrary $y \in D$.

THEOREM 4.1. _Suppose that_ $D \subset D_1$. _Then_

(4.2) $$\lim_{t\to 0} G_D(y, te)/t$$

exists. The limit is greater than 0 _if and only if_ $A = D_1 \setminus D$ _is minimal thin in_ D_1 _at_ $0 \in L$.

PROOF. For some c_1, $0 < c_1 < \infty$ the closure (in R^n) of the set

$$B = \{x \in D : G_D(y, x) > c_1\}$$

is compact and contained in D. Then for all $x \in D \setminus B$

$$P_D^x(T_B < \infty) = G_D(y, x)/c_1.$$

If A is minimal thin in D_1 at 0 then by Lemma 3.1 of Part I and Theorem 4.4 of Burdzy (1984)

$$\lim_{t\to 0} P_D^{te}(T_B < \infty)/t = \lim_{t\to 0} G_D(y, te)/(tc_1)$$

exists and is greater than 0. Thus if the limit (4.2) is zero then A is not minimal thin in D_1 at 0.

If the limit (4.2) exists and is greater than zero then by Lemma 3.1 of Part I and Theorem 4.5 of Burdzy (1984) A is minimal thin in D_1 at 0.

It remains to prove that if A is not minimal thin then the limit (4.2) exists and is equal to zero. Let

$$D_k = D \cup (D_1 \cap \{|x| < 1/k\}) \text{ for } k > 2.$$

Choose c_2, $0 < c_2 < \infty$ so large that

$$\{x \in D_2 : G_{D_2}(y,x) > c_2\} \subset B.$$

If

$$B_k = \{x \in D_k : G_{D_k}(y,x) > c_2\}, \; k > 2$$

then

$$B_j \subset B_k \text{ for } j > k > 2$$

and for $k > 2$

$$\begin{aligned} G_{D_k}(y,te)/c_2 &= P^{te}_{D_k}(T(B_k) < \infty) \\ &= P^{te}_{D_2}(T(B_k) < T(\partial D_k)) \\ &< P^{te}_{D_2}(T_B < T(\partial D_k)). \end{aligned}$$

Let H^0 be the standard excursion law in D_2. Then Theorem 4.4 of Burdzy (1984) implies that for some c_3

$$c_3 H^0(T_B < T(\partial D_k)) = \lim_{t\to 0} P_{D_2}^{te}(T_B < T(\partial D_k))/t$$

$$\geq \lim_{t\to 0} G_{D_k}(y, te)/(tc_2)$$

$$\geq \limsup_{t\to 0} G_D(y, te)/(tc_2).$$

One has $H^0(T_B < \infty) < \infty$ and

$$\{T_B < T(\partial D_k)\} \downarrow \{T_B < T_{\partial D}\}.$$

If A is not minimal thin at 0 then by Lemma 3.1 $H^0(T_B < T_{\partial D}) = 0$ and by continuity of H^0

$$\lim_{k\to\infty} H^0(T_B < T(\partial D_k)) = 0.$$

Therefore

$$\limsup_{t\to 0} G_D(y, te)/t = 0.$$

REMARK 4.1. If the cone condition (4.1) were not assumed then the nonvanishing of the limit (4.2) would depend on the local behavior of the set A near the line $\{x = te, t > 0\}$, which seems more like a nuisance than a real part of the problem. In fact the above theorem is true without assumption (4.1) if the limit in (4.2) is taken not along a line but in the minimal fine topology. In such a case it is not necessary to assume that D_1 is a halfspace.□

Suppose that

$$D = \{x \in D_1 : x_1 > h(\mathrm{proj}_L x)\}$$

for some nonnegative function $h : L \to R$ such that $h(0) = 0$.

COROLLARY 4.1. *If* h *is Lipschitz or* $h(x) = h_1(|x|)$ *for some monotone function* $h_1 : [0,\infty) \to R$ *then the limit*

$$\lim_{t\to 0} G_D(y,te)/t$$

is greater than zero if and only if

$$\int_{L\cap\{|x|<1\}} \frac{h(x)}{|x|^n}\, dx < \infty.$$

PROOF. Use Theorems 3.2 and 4.1.

REMARK 4.2. It is easy to see that whenever $D \subset D_1$ and (4.1) holds then

$$\limsup_{t\to 0} G_D(y,te)/t < \infty$$

without any further assumptions. □

Suppose D is a region such that $0 \in \partial D$. Fix an arbitrary point $y \in D_1$ and a compact nonpolar set $B \subset D_1$.

THEOREM 4.2. *If* $D_1 \subset D$ *then the limit*

$$\lim_{t\to 0} G_D(y,te)/t$$

exists (finite or infinite). The limit is finite if and only if

(4.3) $$\int_L P_D^x(T_B < \infty)|x|^{-n}dx < \infty. \square$$

REMARK 4.3. Choose a constant $c < \infty$ so large that the closure of

$$B_1 = \{x \in R^n : G_D(y,x) > c\}$$

is compact and contained in D_1. Then for $x \in D_1 \setminus B_1$

$$P_D^x(T_{B_1} < \infty) = G_D(y,x)/c$$

and (4.3) is therefore equivalent to a purely potential theoretic condition

$$\int_L \frac{G_D(y,x)}{|x|^n} dx < \infty.$$

PROOF. Let H_1^0 be the standard non-null excursion law in D_1.

If the H_1^0-excursions are continued after their lifetime R as an independent BM_D starting from X(R−) then the distribution H^0 of the resulting process is an excursion law in D (see Burdzy (1984)).

The density of the distribution of $X(T_{L-})$ is proportional to $|x|^{-n}$ for $x \in L$ under the law H_1^0 and therefore also under the law H^0 (see Theorem 3.3 iv) in Burdzy (1984)).

By the strong Markov property applied at T_L

$$(4.4) \quad H^0(T_B < \infty) = H_1^0(T_B < \infty) + c_1 \int_L P_D^x(T_B < \infty)|x|^{-n}dx.$$

By Theorem 3.3 iii) of Burdzy (1984)

$$H_1^0(T_B \ \infty) < \infty.$$

Thus if (4.3) holds then $H^0(T_B < \infty) < \infty$, i.e. H^0 is standard.

Theorem 4.4 of Burdzy (1984) implies that

$$(4.5) \qquad \lim_{t\to 0} P_D^{te}(T_B < \infty)/t = c_2 H^0(T_B < \infty)$$

and the limit is finite. If B is chosen as B_1 in Remark 4.3 then

$$(4.6) \quad \lim_{t\to 0} G_D(y, te)/t = \lim_{t\to 0} P_D^{te}(T_B < R)\cdot c/t < \infty.$$

Assume now that the limit (4.3) exists and is finite. Then by Theorem 4.5 of Burdzy (1984) H^0 is standard and in particular $H^0(T_B < \infty) < \infty$. Therefore it follows from (4.4) that (4.3) holds.

It remains to show that the limit (4.2) exists even if (4.3) does not hold.

For k > 2 define

$$D_k = D_1 \cup (D \cap \{x > 1/k\}).$$

It is easy to see that (4.3) holds if D is replaced by a set D_k. Combine (4.4), (4.5) and (4.6) to obtain for all $k \geq 2$ and the set B defined as B_1 in Remark 4.3

$$\lim_{t\to 0} G_{D_k}(y,te)/t = c_3 H_1^0(T_B < \infty)$$

$$+ c_4 \int_L P_{D_k}^x(T_B < R)|x|^{-n}dx.$$

It is easy to see that the last integral grows to infinity when $k \to \infty$ provided

$$\int_L P_D^x(T_B < R)|x|^{-n}dx = \infty.$$

In such a case

$$\liminf_{t\to 0} G_D(y,te)/t \geq \lim_{k\to\infty}\lim_{t\to 0} G_{D_k}(y,te)/t = \infty. \square$$

Now suppose that $h : L \to R$ is nonnegative and $h(0) = 0$. Let

$$D = \{x \in R^n : x_1 > -h(\mathrm{proj}_L x)\}.$$

Fix $y \in D_1$.

COROLLARY 4.2. <u>Suppose</u> h <u>is Lipschitz. Then</u>

(4.7) $$\lim_{t\to 0} G_D(y,te)/t < \infty$$

<u>if and only if</u>

(4.8)
$$\int_{L\cap\{|x|<1\}} \frac{h(x)}{|x|^n}\,dx < \infty. \square$$

The proof of the above corollary is long and will be preceded by a few lemmas.

Let A_1, A_2, B_1, B_2 be closed sets such that

$$A_1,\ B_1 \subset \{x_1 > 0\},$$

$$A_2,\ B_2 \subset \{x_1 < 0\},$$

$$B_1 = \{x \in R^n : (x_1, \ldots, x_n) = (-y_1, y_2, \ldots, y_n)$$

for some $y \in B_2\}$,

$$A_1 \subset \{x \in R^n : (x_1, \ldots, x_n) = (-y_1, y_2, \ldots, y_n)$$

for some $y \in A_2\} = A_3$,

$$D_2 = R^n \setminus (A_1 \cup A_2).$$

LEMMA 4.1. $P^0_{D_2}(T(B_1-) < \infty) > P^0_{D_2}(T(B_2-) < \infty)$.

PROOF. Let $T_0 = 0$,

$$T^0_k = \inf\{t > T_{k-1} : X(t) \in A_3\} \text{ for } k > 1,$$

$$T_k = \inf\{t > T^0_k : X(t) \in L\} \text{ for } k > 1.$$

By symmetry of Brownian motion one has for all $x \in L$

$$P^x_{D_2}(T_{B_1-} < T_{A_2 \cup A_3 -}) = P^x_{D_2}(T_{B_2-} < T_{A_2 \cup A_3 -}).$$

By the above equality and the strong Markov property at T_k

$$P^0_{D_2}(T_k < T_{B_1-} < T_{k+1},\ X(t) \notin A_2 \cup A_3$$

$$\text{for all } t \in (T_k, T_{B_1-})).$$

$$= P^0_{D_2}(T_k < T_{B_2-} < T_{k+1},\ X(t) \notin A_2 \cup A_3$$

$$\text{for all } t \in (T_k, T_{B_2-})).$$

Note that the event $\{T(B_2-) < \infty\}$ is the disjoint union from $k = 0$ to ∞ of the events on the right hand side of the last inequality. The events on the left hand side are also disjoint for distinct k's and their union is contained in $\{T(B_1-) < \infty\}$. Therefore

$$P^0_{D_2}(T_{B_1-} < \infty) > P^0_{D_2}(T_{B_2-} < \infty). \square$$

Suppose $h : L \to R$ is a nonnegative Lipschitz function, such that $h(0) > 0$. Let $a > 0$ and

$$D_4 = \{x \in R^n : x_1 > -h(\text{proj}_L x) \text{ and } |x_k| < a$$

$$\text{for all } k = 1,2,\ldots,n\}.$$

LEMMA 4.2. *There exist constants* c_1, $c_2 > 0$ *such that if* $h < c_1 a$,

$$A = \{x \in \partial D_4 : \max\{|x_k| : k = 1,\ldots,n\} = a\}$$

and

$$B = \{x : x_1 = c_1 a\}$$

then

$$P^0_{D_4}(T_B < \infty) > c_2 P^0_{D_4}(T_{A-} < \infty).$$

PROOF. Define

$$K_j = \{x \in R^n : x_1 = c_3 x_{j+1}\} \text{ for } j = 1,\ldots,n-1,$$

$$K_j = \{x \in R^n : x_1 = -c_3 x_{j-(n-2)}\} \text{ for } j = n,\ldots,2(n-1).$$

For a set E let $\mathrm{Sym}_j(E)$ be the set symmetric to E with respect to K_j. Let $V_j^1(V_j^2)$ be the closure of the part of $R^n \setminus (D_4 \cup K_j)$ which lies below (above) K_j. Easy geometry shows that if c_3 and c_1 are sufficiently small, $c_1 < c_3$ and $h < c_1 \cdot a$ then

(4.9) $\quad V_j^2 \subset \mathrm{Sym}_j(V_j^1)$ for all $j = 1,\ldots,2(n-1)$.

Let $W_j = V_j^1 \cap \partial D_4 \cap \{|x_j| = a\}$. Then by (4.9) and Lemma 4.1

$$(4.10)\qquad P^0_{D_4}(T_{\mathrm{Sym}_j(W_j)} < \infty) > P^0_{D_4}(T_{W_j-} < \infty).$$

One has

$$(4.11)\qquad \{T_{\mathrm{Sym}_j(W_j)} < \infty\} \subset \{T_B < \infty\}$$

and if

$$W = A \setminus \bigcup_{j=1}^{2(n-1)} W_j$$

then

$$(4.12)\qquad \{T_{W-} < \infty\} \subset \{T_B < \infty\}.$$

One also has

$$\{T_{A-} < \infty\} = \{T_{W-} < \infty\} \cup \bigcup_{j=1}^{2(n-1)} \{T_{W_j-} < \infty\}$$

and the last union is disjoint $P^0_{D_4}$-a.s. Therefore at least one of the events on the right hand side of the last formula must have the $P^0_{D_4}$-probability not less than

$$\frac{1}{2(n-1)+1} P^0_{D_4}(T_{A-} < \infty) = b.$$

If $P^0_{D_4}(T_{W-} < \infty) > b$ then by (4.12)

$$P^0_{D_4}(T_B < \infty) > \frac{1}{2(n-1)+1} P^0_{D_4}(T_{A-} < \infty).$$

If $P^0_{D_4}(T_{W_j-} < \infty) > b$ for some j then the same inequality holds by (4.10) and (4.11) and this completes the proof. □

Let $h : L \to R$ be a Lipschitz and nonnegative function. Assume that $h(x) = 0$ for $|x| < 1/8$ and $|x| > 4$. Let

$$D_2 = \{x \in R^n : -h(\mathrm{proj}_L x) < x_1 < 1 \text{ and } 1/16 < |x| < 8\}.$$

Let μ be the surface area measure on L restricted to $\{1/2 < |x| < 1\}$. Thus μ has finite mass. Let $M(a)$ denote $\{x : x_1 = a\}$ for real a. Denote

$$V = \partial D_2 \setminus \{x : x_1 < 1 \text{ and } 1/4 < |\mathrm{proj}_L x| < 2\}.$$

LEMMA 4.3. <u>There exist constants</u> $c_1, c_2 > 0$ <u>such that if</u> $h < c_1$ <u>then</u>

$$P^{\mu}_{D_2}(X(R-) \in V) < c_2 \int_L h(x)dx.$$

PROOF. Assume that $h < 1$. For each integer $k > 1$ let Q^k_1, Q^k_2, ... be the sequence of all sets of the form

$$\left\{x \in R^n : k_2 2^{-k} < x_2 < (k_2 + 1)2^{-k}, \ldots, k_n 2^{-k} < x_n < (k_n + 1)2^{-k}\right\}$$

where $k_2, \ldots, k_n$ are arbitrary integers. Define inductively some classes of sets.

A set $A \subset R^n$ is a set of class 1 if for some Q^1_j

$$A = \{x \in D_2 \setminus D_1 : \text{proj}_L x \in Q_j^1\}$$

and $h(x) > 2^{-1}$ for some $x \in Q_j^1$.

Let E_1 be the union of all the sets of class 1.

A set A is a set of class k if for some Q_j^k

$$A = \{x \in (D_2 \setminus D_1) \setminus \bigcup_{m=1}^{k-1} E_m : \text{proj}_L x \in Q_j^k\}$$

and $h(x) > 2^{-k}$ for some $x \in Q_j^k$. The union of all sets of class k is called E_k.

Let $A_1, A_2, \ldots$ be the sequence of all sets belonging to one of the classes defined above. The class of the set A_k will be denoted $\varkappa(k)$. Define

$$B_k = \text{proj}_L A_k.$$

Let $D_3 = D_1 \cup D_2$ and c_3 be the supremum of $P_{D_3}^{3e}$-density of $X(T_{L-})$. The infimum of $P_{D_3}^{3e}$-density of $X(T(M(2^{-\varkappa(k)+1})))$ on the set $\{|x| < 4\} \cap M(2^{-\varkappa(k)+1})$ is greater than a constant c_4 (independent of k). Then

$$c_4 c_3^{-1} P_{D_3}^{3e}(T(A_k) < T(M(-2^{-\varkappa(k)+1})))$$

$$< P_{D_3}^{3e}(T(A_k + 2^{-\varkappa(k)+1}e) < T(L)).$$

There exist a constant c_5 such that for all k and all $x \in A_k + 2^{-\varkappa(k)+1}e$

$$P_{D_3}^{x}(T(A_k) = T(L) < \infty) > c_5.$$

Then by the strong Markov property at $T(A_k + 2^{-x(k)+1}e)$

$$P^{3e}_{D_3}(T(A_k) = T(L) < \infty) > P^{3e}_{D_3}(T(A_k + 2^{-x(k)+1}e)$$

$$< T(L))\cdot c_5 > c_4\cdot c_3{}^{-1}P^{3e}_{D_3}(T(A_k)$$

$$< T(M(-2^{-x(k)+1})))\cdot c_5.$$

Note that

$$P^{3e}_{D_3}(T(A_k) = T(L) < \infty) < c_3\cdot \text{area}(B_k)$$

$$< c_3\cdot 2^{-x(k)(n-1)}$$

and therefore

$$P^{3e}_{D_3}(T(A_k) < T(M(-2^{-x(k)+1}))) < c_4{}^{-1}c_3c_5{}^{-1}c_3 2^{-x(k)(n-1)}$$

$$= c_6 2^{-x(k)(n-1)}.$$

The measure μ is absolutely continuous with respect to the $P^{3e}_{D_3}$-distribution of $X(T_{L-})$ and the Radon-Nikodym derivative is bounded away from ∞. Therefore

$$(4.13)\quad P^{\mu}_{D_2}(T(A_k) < T(M(-2^{-x(k)+1}))) < P^{\mu}_{D_3}(T(A_k)$$

$$< T(M(-2^{-x(k)+1}))) < c_7 2^{-x(k)(n-1)}.$$

Denote $S(x,r) = \{z : |x - z| = r\}$. It follows easily from Theorem 4 of Dahlberg (1977) that there exists a constant c_8 such that

$$P^{te}(T(S(0,1/32)) < T_L) < c_8 t$$

for all $0 < t < 1/64$. Assume now that $h < 1/64$. Let $W = \partial D_2 \cap D_1$. Then for all k, $x \in A_k$ and $\tilde{x} = \text{proj}_{M(2^{-\varkappa(k)+1})} x$

$$P^x_{D_2}(T_{W-} < \infty = T(M(-2^{-\varkappa(k)+1}))) < P^x(T(S(\tilde{x}, 1/32))$$

$$< T(M(-2^{-\varkappa(k)+1}))) < c_8 \cdot 2^{-\varkappa(k)+1}.$$

By the strong Markov property at $T(A_k)$ and (4.13)

$$(4.14) \quad P^{\mu}_{D_2}(T(A_k) < T_{W-} < \infty = T(M(-2^{-\varkappa(k)+1})))$$

$$< c_7 2^{-\varkappa(k)(n-1)} \cdot c_8 2^{-\varkappa(k)+1}$$

$$= c_9 \cdot 2^{-\varkappa(k)n} < c_{10} \int_{B_k} h(x)dx.$$

The last inequality follows easily from the Lipschitz property of h.

Define a random number $\tilde{k}$. If a path hits $M(-2^{-k_0})$ but does not hit $M(-2^{-k_0+1})$ then $\tilde{k}$ is defined as the minimum of k's such that

$$\varkappa(k) = \varkappa(k_0) \text{ and } T(A_k) < \infty.$$

Then by (4.14)

$$P^{\mu}_{D_2}(X(R-) \in W) = P^{\mu}_{D_2}\left(\bigcup_{k=1}^{\infty} \{\tilde{k} = k \text{ and } T_{W-} < \infty\}\right)$$

$$= \sum_{k=1}^{\infty} P^{\mu}_{D_2}(\tilde{k} = k \text{ and } T_{W-} < \infty)$$

$$< \sum_{k=1}^{\infty} P^{\mu}_{D_2}(T(A_k) < T_{W-} < \infty = T(M(-2^{-x(k)+1})))$$

$$< \sum_{k=1}^{\infty} c_{10} \int_{B_k} h(x)dx$$

$$= c_{10} \int_{L} h(x)dx.$$

Now Lemma 4.2 will be applied with $a = 1/8$. Suppose c_{11} and c_{12} satisfy Lemma 4.2 (in place of c_1 and c_2). Let $D(x)$, $A(x)$ and $B(x)$ be the sets D_4, A and B of Lemma 4.2 translated by x.

Assume that $h < c_{11} \cdot 1/8$.

For every $x \in B(x)$

$$P^{x}_{D_2}(X(R-) \in W) > P^{x}(T(M(1)) < T(L))$$

$$= c_{11} 1/8.$$

Thus by the strong Markov property at $T_{B(x)}$

$$P^{\mu}_{D_2}(X(R-) \in W) = \int P^{x}_{D_2}(X(R-) \in W)d\mu(x)$$

$$> c_{11} \cdot 1/8 \cdot \int P^{x}_{D_2}(T_{B(x)} < \infty)d\mu(x).$$

This and Lemma 4.2 imply that

$$P^{\mu}_{D_2}(X(R-) \in V) = \int P^{x}_{D_2}(X(R-) \in V)d\mu(x)$$

$$\leq \int P^x_{D(x)}(T_{A(x)-} < \infty)d\mu(x)$$

$$\leq c_{12}^{-1} \int P^x_{D(x)}(T_{B(x)} < \infty)d\mu(x)$$

$$\leq c_{12}^{-1} \int P^x_{D_2}(T_{B(x)} < \infty)d\mu(x)$$

$$\leq c_{12}^{-1}\cdot c_{11}^{-1}\cdot 8\cdot P^{\mu}_{D_2}(X(R-) \in W)$$

$$\leq c_{12}^{-1}\cdot c_{11}^{-1}\cdot 8\cdot c_{10} \int_L h(x)dx. \square$$

PROOF OF COROLLARY 4.2. Suppose that (4.8) does not hold.

Let $c_1 > 0$ be a constant. For every $x \in L$ let $Q(x)$ be the closed cube which has center x, side length $c_1h(x)$ and its sides are parallel to the axes (if $h(x) = 0$ then $Q(x) = \emptyset$). Choose c_1 so that $Q(x) \subset D$ for every x.

Let $Q_1(x) = Q(x) \cap \{x : x_1 = c_1h(x)/2\}$. For all $x \in L$,

$$(4.15) \qquad 0 < P^x_D(T_{Q_1(x)} < T_{\partial Q(x)\setminus Q_1(x)}) = c_2 < 1$$

which follows from the Brownian scaling.

Let $B = D_1 \cap \{|x| < 2\}$ and $B_1 = \partial B \cap \{x_1 > 1\}$. Then for all $t \in (0,1)$

$$(4.16) \qquad P^{te}_B(X(R-) \in B_1) > c_3t,$$

which follows easily from Theorem 4 of Dahlberg (1977).

Let $B_2 = \{x_1 = 1, |x| < 5\}$. B_2 is compact and $B_2 \subset D$. Note that $c_1h(x)/2 < 1$ for $x \in L \cap \{|x| < 1\}$ since

$Q(x) \subset D$. Thus (4.16) and translation invariance of Brownian motion imply that for all $z \in Q_1(x)$ and $\tilde{z} = \text{proj}_L z$

$$(4.17) \qquad P_D^z(T(B_2) < \infty) > P^z_{B+\tilde{z}}(X(R-) \in B_1 + \tilde{z})$$

$$> c_3 \cdot |z - \tilde{z}|$$

$$> c_3 \cdot c_1 h(x)/2.$$

By (4.15), (4.17) and the strong Markov property applied at $T(Q_1(x))$

$$P_D^x(T_{B_2} < \infty) > c_2 c_3 c_1 h(x)/2 > c_4 h(x)$$

for all $x \in L \cap \{|x| < 1\}$. Thus

$$\int_L P_D^x(T_{B_2} < R)|x|^{-n} dx > \int_{L \cap \{|x|<1\}} \frac{c_4 h(x)}{|x|^n} dx = \infty$$

and (4.7) follows from Theorem 4.2.

Now suppose that (4.8) is satisfied.

Note that by Theorem 4 of Dahlberg (1977) the finiteness of the limit in (4.7) depends only on the local behavior of h near 0. Conditions imposed without loss of generality (WLOG) below refer to this remark.

It follows from (4.8) that $h(x)/|x| \to 0$ when $|x| \to 0$. Assume WLOG that $h(x)/|x| < c_1/4$ for all x, where c_1 is the constant of Lemma 4.3. If

$$\nu(1)(dx) = \nu_1(dx) = |x|^{-n} 1_{\{1/2<|x|<1\}} dx$$

then $\nu_1(dx) < \mu(dx)\cdot 2^n$, where μ is defined in Lemma 4.3.

Let $h_2 : L \to R$ be the smallest Lipschitz nonnegative function (with the same constant as h) which majorizes h on $\{1/4 < |x| < 2\}$. Assume WLOG that $h(x)/|x| < c_3$ for all x and c_3 is so small that $h_2(x) = 0$ for $|x| < 1/8$ and $|x| > 4$. Let D_2 be the region defined in Lemma 4.3, relative to h_2. By Lemma 4.3,

$$P_D^{\nu(1)}(T_{\partial D_2} < \infty) < 2^n c_2 \int_L h_2(x)dx$$

$$< c_4 \int_{L\cap\{1/8<|x|<4\}} h(x)dx.$$

Let M(a) have the same meaning as in Lemma 4.3. There exists a constant c_5 such that for all $x \in \partial D_2$ the P_D^x-density of $X(T_{M(2)})$ is less than $c_5/(1 + |\text{proj}_L y|^n)$ at the point $y \in M(2)$, which follows from Proposition 2.2.22 of Port and Stone (1978). It follows that the $P_D^{\nu(1)}$ density of $X(T_{M(2)})$ is less than

(4.18) $$(1 + |\text{proj}_L y|^n)^{-1} c_4 c_5 \int_{L\cap\{1/8<|x|<4\}} h(x)dx$$

at the point $y \in M(2)$.

For $y \in M(a)$, $0 < a < 3$ one has

(4.19) $$P^y(T_{M(4)} < T_L) < c_6 a.$$

Note also that

(4.20) $$\int_L (1 + |x|^n)^{-1} dx < c_7 < \infty.$$

Define

$$T_0 = 0,$$

if $|X(T_{k-1})| \in (2^{-j-2}, 2^{-j-1}]$ then $S_k = \inf\{t > T_{k-1} : X(t) \in M(2^{-j})\}$, $k > 1$,

$$T_k = \inf\{t > S_k : X(t) \in L\}, \ k > 1.$$

By the strong Markov property at $T_{M(2)}$, (4.18), (4.19) and (4.20)

$$P_D^{\nu(1)}(T_{M(4)} < T_1) < c_4 \cdot c_5 \int_{L \cap \{1/8 < |x| < 4\}} h(x) dx \cdot c_7 \cdot c_6 \cdot 2$$

$$= c_8 \int_{L \cap \{1/8 < |x| < 4\}} h(x) dx.$$

Assume WLOG that $h(x) = 0$ for $|x| > 1$. Using Brownian scaling one obtains analogously for

$$\nu(k)(dx) = \nu_k(dx) = |x|^{-n} 1_{\{2^{-k} < |x| < 2^{-k+1}\}} dx, \ k > 1$$

that

$$P_D^{\nu(k)}(T_{M(4)} < T_1) < 2^{(k-1)n} c_8 \int_{L \cap \{2^{-k-2} < |x| < 2^{-k+3}\}} h(x) dx$$

$$< c_9 \int_{L \cap \{2^{-k-2} < |x| < 2^{-k+3}\}} \frac{h(x)}{|x|^n} dx.$$

If $\nu = \sum_{k=1}^{\infty}\nu_k$ then

$$P_D^{\nu}(T_{M(4)} < T_1) < c_9 \sum_{k=1}^{\infty} \int_{L\cap\{2^{-k-2}<|x|<2^{-k+3}\}} \frac{h(x)}{|x|^n}\,dx$$

$$< 5\cdot c_9 \int_L \frac{h(x)}{|x|^n}\,dx$$

$$= c_{10} \int_L \frac{h(x)}{|x|^n}\,dx.$$

By Brownian scaling, the $P_D^{\nu(k)}$-density of $X(T(M(2^{-k+2})))$ is less than

$$2^{2(k-1)n} c_{11} \int_{L\cap\{2^{-k-2}<|x|<2^{-k+3}\}} h(x)dx$$

$$\times\ (1 + |\text{proj}_L y|^n 2^{(k-1)n})^{-1}$$

at $y \in M(2^{-k+2})$ (see (4.14)). If $\eta(dx) =$ $(1 + |\text{proj}_L x|^n)^{-1}dx$ is a distribution on $M(b)$ for some $0 < b < 3$ then the P^{η}-density of $X(T_L)$ is less than $c_{12}(1 + |x|^n)^{-1}$ at $x \in L$ which follows from the strong Markov property and Proposition 2.2.22 of Port and Stone (1978). Thus the $P_D^{\nu(k)}$ -density of $X(T_1)$ at the point $x \in L$ is less than

$$2^{2(k-1)n} c_{13} \int_{L\cap\{2^{-k-2}<|x|<2^{-k+3}\}} h(x)dx\cdot(1 + |x|^n 2^{(k-1)n})^{-1}$$

$$< 2^{(k-1)n} c_{13} \int_{L\cap\{2^{-k-2}<|x|<2^{-k+3}\}} h(x)dx\cdot\frac{2^{(k-1)n}}{1 + |x|^n 2^{(k-1)n}}$$

$$< c_{14} \int_{L\cap\{2^{-k-2}<|x|<2^{-k+3}\}} \frac{h(x)}{|x|^n}\,dx\cdot\frac{1}{2^{-(k-1)n} + |x|^n}$$

$$< c_{15} \int_{L\cap\{2^{-k-2}<|x|<2^{-k+3}\}} \frac{h(x)}{|x|^n}dx\cdot\frac{1}{|x|^n}.$$

Choose c_{15} so that the last inequality holds for all $|x| < 1$. The P_D^ν-density of $X(T_1)$ is less than

$$c_{15} \sum_{k=1}^{\infty} \int_{L \cap \{2^{-k-2} < |x| < 2^{-k+3}\}} \frac{h(x)}{|x|^n} dx \cdot \frac{1}{|x|^n}$$

$$< c_{15} \cdot 5 \int_L \frac{h(x)}{|x|^n} dx \cdot \frac{1}{|x|^n}.$$

Now assume WLOG that

$$5c_{15} \int_L \frac{h(x)}{|x|^n} \, dx < c_{16} < 1.$$

By the strong Markov property at T_j and the induction argument

$$P_D^\nu(T_j < T_{M(4)} < T_{j+1}) < c_{16}^j c_{10} \int_L \frac{h(x)}{|x|^n} \, dx$$

and P_D^ν-density of $X(T_j)$ is less than $c_{16}^j |x|^{-n}$. By continuity of paths

$$\{T_{M(4)} < \infty\} \subset \bigcup_{k=1}^{\infty} \{T_{k-1} < T_{M(4)} < T_k\} \ P_D^\nu\text{-a.s.}$$

and

$$P_D^\nu(T_{M(4)} < \infty) < \sum_{j=1}^{\infty} c_{16}^{j-1} c_{10} \int_L \frac{h(x)}{|x|^n} \, dx < \infty.$$

Now (4.7) follows from Theorem 4.2. □

Suppose that $h : L \to R$ is given by $h(x) = h_1(|x|)$ for some nonnegative monotone function $h_1 : [0,\infty) \to R$. Let

$$D = \{x \in R^n : x_1 > -h(\mathrm{proj}_L x)\}.$$

Fix a point $y \in D$.

COROLLARY 4.3. *The limit*

$$\lim_{t \to 0} G_D(y, te)/t$$

is finite if and only if

$$\int_{L \cap \{|x| < 1\}} \frac{h(x)}{|x|^n} \, dx < \infty. \square \tag{4.21}$$

PROOF. The proof is completely analogous to the proof of Theorem 2.4 and therefore is omitted. The function h lies between two Lipschitz functions which satisfy or do not satisfy (4.21) simultaneously. □

Let $h : L \to R$ be a nonnegative function such that $h(0) = 0$. Suppose that $D \subset R^n$ is such a region that for some neighborhood U of 0 and all $x \in \partial D \cap U$

$$|x_1| < h(\mathrm{proj}_L x).$$

Fix $y \in D$.

COROLLARY 4.4. *Assume that* h *is Lipschitz or* $h(x) = h_1(|x|)$ *for some monotone function* $h_1 : [0,\infty) \to R$. *If*

$$\int_{L \cap U} \frac{h(x)}{|x|^n} \, dx < \infty \tag{4.22}$$

then

$$\text{(4.23)} \qquad \lim_{t\to 0} G_D(y,te)/t$$

exists and is different from 0 and ∞. □

PROOF. The existence follows from Theorem 4.4 of Burdzy (1984). Note that the assumption (D) of that theorem is satisfied due to Corollaries 4.1, 4.2 and 4.3.

The Green function is a monotone function of the region, i.e. if $D_2 \subset D_3$ and $x,y \in D_2$ then

$$G_{D_2}(x,y) < G_{D_3}(x,y).$$

Therefore the limit (4.23) is different from 0 and ∞ by comparison with regions considered in Corollaries 4.1, 4.2 and 4.3. □

REMARKS 4.2. i) The condition (4.22) is not necessary. Although this observation seems trivial at the first glance (and it is trivial!) its deeper meaning is revealed by the following example.

Suppose $n = 2$. There exists a Lipschitz function $h : L \to R$ which is also monotone (in the obvious sense) and such that $h(0) = 0$ and

$$\int_{L\cap\{0<x_2<1\}} \frac{|h(x)|}{|x|^2}\,dx = \int_{L\cap\{-1<x_2<0\}} \frac{|h(x)|}{|x|^2}\,dx = \infty.$$

Despite the above the limit (4.23) exists and is different from 0 and ∞ for the region D defined by

$$D = \{x \in R^2 : x_1 > h(\text{proj}_L x)\}.$$

See Section 6 of Burdzy (1985b) for more information about this example.

ii) The Corollaries 4.2 and 4.3 amount to finding criteria for minimal thinness of the set $D \setminus D_1$ in D at the point 0. This remark however is not so useful as the analogous observation in the case $D \subset D_1$.

iii) The above corollaries prove and substantially extend the conjecture following Theorem 4.3 in Burdzy (1984).

iv) Widman (1967) showed that the integral condition (4.22) implies existence, finiteness and nonvanishing of the limit (4.23). His conditions imposed on h were much more restrictive than the ones given in this paper. The above theorems concerning the Green function will be applied in Burdzy (1985b) to the angular derivative problem. The results of Widman (1967) are not strong enough for this purpose.

References

[1] Burdzy, K. (1984) "Brownian excursions from hyperplanes and smooth surfaces" - a preprint.

[2] Burdzy, K. (1985a) "Brownian excursions and minimal thinness. Part I." - a preprint.

[3] Burdzy, K. (1985b) "Brownian excursions and minimal thinness. Part III. Applications to the angular derivative problem." - a preprint.

[4] Dahlberg, B. E. J. (1977) "Estimates of harmonic measure." Arch. Rat. Mech. 65, 275-288.

[5] Jerison, D. S. and Kenig, C. E. (1982) "Boundary behavior of harmonic functions in non-tangentially accessible domains." Adv. in Math. 46, 80-147.

[6] Jerison, D. S. and Kenig, C. E. (1983) "Boundary value problem in Lipschitz domains." Studies in Partial Differential Equations. Walter Littman, Ed., MAA Studies in Mathematics 23.

[7] Port, S. C. and Stone, C. J. (1978) Brownian Motion and Classical Potential Theory. Academic Press, New York.

[8] Widman, K.-O. (1967) "Inequalities for the Green function and boundary continuity of the gradient of solutions of elliptic differential equations." Math. Scand. 21, 17-37.

[9] Ziemer, W. P. (1974) "Some remarks on harmonic measure in space." Pacific J. Math. 55, 629-637.

Krzysztof Burdzy
Department of Mathematics
University of California, San Diego
La Jolla, California 92093

Seminar on Stochastic Processes, 1985
Birkhäuser, Boston, 1986

DOUBLY-FELLER PROCESS WITH MULTIPLICATIVE FUNCTIONAL

by

K. L Chung*

Despite the common use of the term "Feller property", there are variations in its definition. In the early literature on Markov processes, there are discussions of this and related properties, often under sets of bewildering assumptions. The coast should now be clear, but certain neat formulations may have been overlooked. In §1 of this note, some old results are reviewed in more general forms and an apparently new one is derived. In §2, the results are extended to include a multiplicative functional, of which the prime example is that of Feynman-Kac, properly generalized.

1. Doubly-Feller Process

Let E be a locally compact separable metric space, $E_\Delta = E \cup \{\Delta\}$ its one-point compactification, $\mathscr{E}$ its Borel

*This work was supported in part by NSF grant DMS83-01072 at Stanford University.

tribe. Any function f defined on E is extended to E_Δ by setting $f(\Delta) = 0$.

$b\mathscr{E}$ = the class of bounded $\mathscr{E}$-measurable functions on E;

bC = the class of bounded continuous functions on E;

C_0 = the class of continuous functions on E such that its extension to E_Δ (as specified above) is continuous in E_Δ.

For $f \in b\mathscr{E}$, we write $\|f\| = \sup\{|f(x)| : x \in E\}$,

Let $\{P_t, t > 0\}$ be a probability transition semigroup with state space E_Δ; P_0 being the identity. It is said to have the Feller property if the following two conditions are satisfied:

(i) for each $t > 0$ and $f \in C_0$, we have $P_t f \in C_0$;

(ii) for each $f \in C_0$, $\lim_{t \downarrow 0} P_t f(x) = f(x)$.

It is known that (i) and (ii) together imply that for each $f \in C_0$,

(ii') $$\lim_{t \downarrow 0} \|P_t f - f\| = 0.$$

The semigroup (P_t) is said to have the strong Feller property if

(iii) for each $t > 0$ and $f \in b\mathscr{E}$, $P_t f \in bC$.

It is well known ([1], Chapter 2) that if (P_t) has the

Feller property, then a Hunt process $X = \{X_t, t > 0\}$ can be constructed with (P_t) as its transition semigroup. Thus the paths are right continuous and quasi left continuous. Such a process will be called a Feller process. Let $\{\mathscr{F}_t^\sim, t > 0\}$ be the augmented natural filtration. For each $B \in \mathscr{E}$, define

$$T_B = \inf\{t > 0 : X_t \in B\};$$

$$\tau_B = T_{E-B}.$$

For a Hunt process, it is known that for each $t > 0$, $\{T_B < t\} \in \mathscr{F}_t^\sim$, and the function $x \to P^x\{T_B < t\}$ is universally measurable, denoted by $\mathscr{E}^\sim$. A full discussion of these questions of measurability is given in [1], §2.3. We begin with a useful consequence.

LEMMA 1. _If_ (P_t) _has the strong Feller property, then for each_ $t > 0$ _and_ $f \in b\mathscr{E}^\sim$ _(bounded and in_ $\mathscr{E}^\sim$_), we have_ $P_t f \in bC$.

PROOF. (see [3], Annexe 5): Let $\{x_n\}$ be a dense set in E, and

$$\lambda = \sum_{n=1}^{\infty} 2^{-n} \varepsilon_{x_n}$$

where ε_x is the point mass at x. Given $f \in b\mathscr{E}^\sim$, there exist f_1 and f_2 in $\mathscr{E}$ such that $f_1 < f < f_2$ and $(\lambda P_t)(f_1) = (\lambda P_t)(f_2)$. Hence $P_t f_1 < P_t f < P_t f_2$ and $\lambda(P_t f_2 - P_t f_1) = 0$. Since $P_t f_1$ and $P_t f_2$ are continuous and the measure λ charges every nonempty open set, it follows that $P_t f_1 = P_t f_2$, and consequently $P_t f = P_t f_1 \in bC$.

The following result is proved in [2] (or see [1], p. 73, Exercise 2).

LEMMA 2. *Let* X *be a Feller process. For each nonempty open set* B *and compact subset* K *of* B, *we have*

$$(1) \qquad \lim_{t \to 0} \sup_{x \in K} P^x\{\tau_B \le t\} = 0.$$

REMARK. A condition like (1) can be found in [4] under assumption of continuous paths, but it was not deduced from the "Feller property" which was defined differently there.

The next result is given for the Brownian motion and an open set B in at least four textbooks, including [1]. Since the general case is not stated, we repeat its known proof here to illustrate the measurability question.

LEMMA 3. *Let* X *be a Hunt process whose transition semigroup* (P_t) *has the strong Feller property. Then for each* $B \in \mathcal{E}$, *both functions below are upper semi-continuous:*

$$(2) \qquad x \to P^x\{t < T_B\}, \qquad x \to P^x\{t \le T_B\}.$$

PROOF. We have the fundamental relation:

$$T_B = \lim_{s \downarrow 0} \downarrow (s + T_B \circ \theta_s)$$

where θ_s is the shift operator. It follows that

$$(3) \quad P^x\{t < T_B\} = \lim_{s \downarrow 0} \downarrow P^x\{t - s < T_B \circ \theta_s\}$$

$$= \lim_{s \downarrow 0} \downarrow P^x\{P^{X_s}[t - s < T_B]\} = \lim_{s \downarrow 0} P_s \phi_s(x)$$

where

$$\phi_s(x) = P^x[t - s < T_B].$$

Hence $\phi_s \in b\mathscr{E}^\sim$ because X is a Hunt Process, and $P_s \phi_s \in bC$ by Lemma 1. It follows from (3) that the first function in (2) is upper semi-continuous. So is the second by the same proof, changing "<" into "<" in the obvious places. □

A **doubly-Feller process** is a Feller process whose transition semigroup has also the strong Feller property. The most famous example is the Brownian motion in R^d. Most diffusion processes are doubly-Feller processes. But we are not assuming continuous paths here.

Let B be a nonempty open subset of E, $B \neq E$; and let $B \cup \{\Delta_B\}$ be its one-point compactification, where $\Delta_B \neq \Delta$. Define

$$(4) \qquad X_t^B = \begin{cases} X_t & \text{on } \{t < \tau_B\}; \\ \Delta_B & \text{on } \{t \geq \tau_B\}; \end{cases}$$

The process $X^B = \{X_t^B, t \geq 0\}$ is called "the process X killed outside B." Its state space is $B \cup \{\Delta_B\}$, and its transition semigroup $\{P_t^B, t \geq 0\}$ is given by

$$(5)\quad \begin{aligned} &P_t^B(x,A) = E^x\{t < \tau_B; X_t \in A\} \text{ if } x \in B,\ A \in B \cap \mathscr{E};\\ &P_t^B(x,\{\Delta_B\}) = 1 - P_t^B(x,E) \quad \text{if } x \in B;\ P_t^B(\Delta_B,\{\Delta_B\}) = 1. \end{aligned}$$

If X is a Hunt process, it can be verified that X^B is also a Hunt process. There is no difficulty with the right continuity and quasi left continuity of paths, while the strong Markov property is shown exactly as in Theorem 2 of §4.5 of [1], which treats the special case where X is the Brownian motion.

We denote by $b\mathscr{E}(B)$, $b\mathscr{E}^\sim(B)$, $bC(B)$ the indicated classes of functions restricted to B. Let B* be the boundary of B in E_Δ. When f is defined only on B, we extend it to E_Δ by setting it to be zero outside B. If this extension is continuous in E_Δ we say that f belongs to $C_0(B)$. The open set B is said to be regular if for each $z \in B^* \cap E$, we have $P^z\{\tau_B = 0\} = 1$. This is the definition used in the Dirichlet problem for B, but note that "regularity at Δ" is not defined when $\Delta \in B^*$.

The following theorem is the main result of this section. It is known when X is the Brownian motion in R^d. In the general form given here it is apparently new, although analogous results may be found in the literature.

THEOREM. *Let* X *be a doubly-Feller process with the state space* E_Δ. *Let* B *be a nonempty proper subset of* E *which is open and regular, and define the process* X^B *by* (4). *Then* X^B *is also a doubly-Feller process*.

PROOF: As remarked above, X^B is a Hunt process with the transition semigroup (P_t^B) given by (5). We prove first that the latter has the strong Feller property. Let $t > 0$, $f \in b\mathscr{E}(B)$. Then by (5),

$$(6) \qquad P_t^B f(x) = E^x\{t < \tau_B;\ f(X_t)\}, \qquad x \in B.$$

For $0 < s < t$, we have by the Markov property:

$$(7) \qquad P_t^B f(x) = E^x\{s < \tau_B;\ \phi_s(X_s)\},$$

where

$$\phi_s(x) = E^x\{t - s < \tau_B;\ f(X_{t-s})\}, \qquad x \in E.$$

Then $\phi_s \in b\mathscr{E}^{\sim}$ and so by Lemma 1, $P_s\phi_s \in bC$. It follows from (7) that

$$(8) \quad \left|P_t^B f(x) - P_s\phi_s(s)\right| < P^x\{\tau_B < s\}\|\phi_s\|, \qquad x \in B.$$

Since $\|\phi_s\| < \|f\|$ for all s, the right member of (8) converges to zero uniformly for x in every compact subset of B by Lemma 2. Hence $P_t^B f \in bC(B)$. This proves the strong Feller property of (P_t^B).

REMARK. If we change "$t < \tau_B$" in the definition of the function in (6) into "$t < \tau_B$", the resulting function is also continuous in B by the same argument, provided that $f \in b\mathscr{E}$.

Next we prove that if $t > 0$ and $f \in C_0(B)$, then

$P_t^B f \in C_0(B)$. Since B* may contain Δ, let us consider first this case. We have then as $x \in B$, $x \to \Delta$:

$$P_t^B|f|(x) \leq P_t|f|(x) \to 0$$

since $|f| \in C_0$, where f is the extension of f to E_Δ, specified above. On the other hand, if $z \in B^* \cap E$, then we have by Lemma 3 and the regularity assumption

$$(9) \qquad \overline{\lim_{x \to z}} P^x\{t < \tau_B\} \leq P^z\{t < \tau_B\} = 0;$$

and consequently as $x \in B$, $x \to z$,

$$|P_t^B f(x)| \leq P^x\{t < \tau_B\}\|f\| \to 0.$$

We have therefore proved that (P_t^B) satisfies condition (i) of the Feller property.

Finally, let $f \in C_0(B)$ and extend it to E_Δ as before. We have then obviously

$$(9) \qquad |P_t^B f(x) - P_t f(x)| \leq P^x\{\tau_B \leq t\}\|f\|.$$

For each $x \in B$, $P^x\{\tau_B = 0\} = 0$ since B is open and the paths are right continuous. Hence it follows from (9) that

$$\lim_{t \downarrow 0} P_t^B f(x) = \lim_{t \downarrow 0} P_t f(x) = f(x), \qquad x \in B$$

by the Feller property of (P_t). Thus (P_t^B) satisfies condition (ii) of the Feller property, and we have concluded the proof that X^B is a doubly-Feller process.

COROLLARY. *For a doubly-Feller process, and a nonempty open set* B, *the two functions in* (2) *are both continuous in* B, *indeed in* E *if* B *is regular*.

PROOF: The first function is just $P_t^B 1_B$. The assertion for the second function is a consequence of the Remark in the preceding proof.

Even in the case of a Brownian motion, the two results in the Corollary seem to have escaped notice in the literature.

2. Multiplicative Functional

Let $M = \{M_t, t > 0\}$ be a multiplicative functional associated with X. Namely, for each $x \in E$, P^x-almost surely: $M_0 \equiv 1$, $0 < M_t < \infty$, $M_t \in \tilde{\mathscr{F}}_t$ for $t > 0$; and

(10) $M_{t+s} = M_s \cdot (M_t \circ \theta_s)$, for all $t > 0$, $s > 0$.

We now impose a set of special conditions on M as follows.

(a) For some $t > 0$:

$$\sup_{x \in E} \sup_{0<s<t} E^x\{M_t\} < \infty.$$

It follows from this, condition (10) and the Markov property that for $x \in E$ and $s \in [0,t]$:

$$E^x\{M_{t+s}\} = E^x\{M_s E^{X_s}[M_t]\} < \sup_{x \in E} \sup_{0<s<t} E^x\{M_s\}^2.$$

Hence by induction (a) is in fact true for all (finite) t. We shall denote the bound by A_t below.

(b) For each $t > 0$, there exists a number $\alpha > 1$ (which may depend on t) such that

$$\sup_{x \in E} E^x\{M_t^\alpha\} < \infty.$$

(c) For each compact subset K of E, we have

$$\lim_{t \to 0} \sup_{x \in K} E^x\{|M_t - 1|\} = 0.$$

These conditions are the simplest to yield Theorem 2 below. They are inspired by our prime example which we take up first.

EXAMPLE. Let $q \in \mathscr{E}$, and suppose that

(11) $$\lim_{t \to 0} \sup_{x \in E} E^x\{\int_0^t |q|(X_s)ds\} = 0.$$

Put

$$e_q(t) = \exp\{\int_0^t q(X_s)ds\}.$$

Then $\{e_q(t), t > 0\}$ is a multiplicative functional satisfying all the conditions above. It is known as the Feynman-Kac functional when X is the Brownian motion. The condition (11) is due to Aizenman and Simon.

It follows from (11), (10) and the Markov property that the sup in (11) is in fact finite for all $t > 0$. The argument is similar to that given under (a), changing multiplication to addition. Hence for each x, P^x-almost surely, we have $0 < e_q(t)$ ∞, and $e_q(t) \in \tilde{\mathcal{F}}_t$. It is easy to verify (10) for $M_t = e_q(t)$. (Indeed $\{\int_0^t q(X_s)ds, t > 0\}$ is an additive functional.) To verify the conditions (a), (b) and (c), let us cite the following lemma due to Khas'minskii [5].

LEMMA 4. <u>If for a fixed t the</u> sup <u>in</u> (11) <u>is less than</u> $\varepsilon < 1$, <u>then</u>

$$(12) \qquad \sup_{x \in E} E^x\{e_{|q|}(t)\} < \frac{1}{1 - \varepsilon}.$$

The proof is by using Taylor's series for the exponential function, then estimating each $E^x\{(\int_0^t |q|(X_s)ds)^n\}$ by converting the nth power into an n - tuple integral, and using the Markov property.

Therefore under (11), condition (a) is satisfied by $M_t = e_{|q|}(t)$ for a sufficiently small $t > 0$ since $e_{|q|}(t)$, is increasing with t. Since $\sup\{e_q(s):0 < s < t\} < e_{|q|}(t)$ this implies condition (a) for $M_t = e_q(t)$. Next, for any real constant α, we have

$$(13) \qquad ((e_q(t))^\alpha = e_{\alpha q}(t) < e_{|\alpha q|}(t).$$

Now (11) remains true when q is replaced by αq; hence condition (b) is satisfied by $M_t = e_q(t)$, indeed for any $\alpha > 1$ independent of t. Finally, let $\varepsilon > 0$ and choose $t > 0$ so that the sup in (12) is less that $1 + \varepsilon$. For this fixed t let $\Lambda = \{e_q(t) > 1\}$. Then for all $x \in E$ we have

(14) $$E^x\{|e_q(t) - 1|\} = E^x\{e_q(t) - 1; \Lambda\}$$

$$+ E^x\{1 - e_q(t); \Lambda^c\}.$$

The first term on the right side is bounded by $E^x\{e_{|q|}(t) - 1; \Lambda\}$. The second term is bounded by

$$E^x\{1 - e_{-|q|}(t); \Lambda^c\} \leq E^x\{e_{|q|}(t) - 1; \Lambda^c\}$$

because $e_{|q|}(t) \geq 1$. Adding up we see the left member in (14) is bounded by $E^x\{e_{|q|}(t)\} - 1 < \varepsilon$. This establishes a stronger form of the condition (c) for $M_t = e_q(t)$.

Now we proceed to extend the results in §1 to a process with multiplicative functional. We begin by defining $\{Q_t, t \geq 0\}$ as follows: for $f \in b\mathscr{E}$:

(15) $$Q_t f(x) = E^x\{M_t f(X_t)\}, \qquad x \in E.$$

By means of (10), we can verify that $\{Q_t\}$ forms a semigroup, not necessarily submarkovian. But we have for each $t \geq 0$,

(16) $$\|Q_t f\| \leq \sup_{x \in E} E^x\{M_t\} \|f\| \leq A_t \|f\|$$

by (a), so that each Q_t maps $b\mathscr{E}$ into $b\mathscr{E}$. The definitions of both Feller properties can be extended to $\{Q_t\}$ without change. We shall keep the notation (P_t) for the semigroup of the process X.

THEOREM 2. If (P_t) has both Feller properties, then so does (Q_t) provided that (M_t) satisfies the conditions (a), (b) and (c).

PROOF: We have by using (10) together with the Markov property, for $x \in E$ and $f \in b\mathscr{E}$:

$$Q_t f(x) = E^x\{M_s \cdot [M_{t-s} f(X_{t-s})] \circ \theta_s\}$$

$$= E^x\{M_s Q_{t-s} f(X_s)\}.$$

Hence

$$(17) \quad |Q_t f(x) - P_s Q_{t-s} f(x)| = |E^x\{(M_s - 1)Q_{t-s} f(X_s)\}|$$

$$\leq E^x\{|M_s - 1|\} \|Q_{t-s} f\| \leq E^x\{|M_s - 1|\} A_t \|f\|$$

by condition (a) and an estimate like (16). Since $Q_{t-s} f \in b\mathscr{E}^{\sim}$, we have $P_s Q_{t-s} f \in bC$ by Lemma 1. Letting $s \to 0$ in (17), the right member converges to zero uniformly in each compact by condition (c). Hence $Q_t f \in bC$, and we have verified the strong Feller property for (Q_t).

Next, if $f \in C_0$, then so is $|f|^\alpha$ for any $\alpha > 0$. We have by Hölder's inequality applied to (15):

$$(18) \qquad |Q_t f(x)| \leq E^x\{M_t^\alpha\}^{1/\alpha} E^x\{|f|^{\alpha'}(X_t)\}^{1/\alpha'}$$

where $\alpha^{-1} + (\alpha')^{-1} = 1$. By condition (b), the first factor on the right side of (18) is bounded in x; as $x \to \Delta$, the second factor converges to zero because (P_t) has the Feller property. Hence $Q_t f \in C_0$.

Finally, if $f \in C_0$, then

$$|Q_t f(x) - P_t f(x)| \leq E^x\{|M_t - 1|\}\|f\|$$

and consequently by a weaker form of condition (c), for each $x \in E$:

$$\lim_{t \to 0} Q_t f(x) = \lim_{t \to 0} P_t f(x) = f(x).$$

Thus (Q_t) satisfies the conditions of the Feller property.□

Combining Theorems 1 and 2, we obtain

THEOREM 3. _Let_ X _be a doubly-Feller process,_ B _as in Theorem_ 1, _and_ M _as in Theorem_ 2. _Define for_ $x \in E$, $f \in b\mathscr{E}$.

(19) $$Q_t^B f(x) = E^x\{t < \tau_B;\ M_t f(X_t)\}.$$

Then $\{Q_t^B, t > 0\}$ _is a doubly-Feller semigroup._

COROLLARY. _If, in addition,_ B _is relatively compact, then we have_ $Q_t^B f \in C_0(B)$ _for each_ $t > 0$ _and_ $f \in b\mathscr{E}(B)$.

PROOF: In this case $\Delta \notin B^*$. Let $z \in B^*$, then we have by (19):

$$|Q_t^B f(x)| \leq E^x\{t < \tau_B\}^{1/\alpha'} E^x\{M_t^\alpha |f|^\alpha (X_t)\}^{1/\alpha}.$$

Hence as $x \to z$, the above converges to zero by (9) and condition (b). Thus $Q_t^B f \in C_0(B)$.

The following alternative approach to Theorem 3 is illuminating. It is well known that the killing operation is representable by a multiplicative functional as follows:

$$\tilde{M}_t = 1_{\{t < \tau_B\}}, \qquad t > 0,$$

provided that the original state space E is replaced by the new state space B, as it should be because X_t lives on B on $\{t < \tau_B\}$. It is easy to verify then all the conditions imposed on $\tilde{M} = \{\tilde{M}_t\}$ at the beginning of this section, starting with the fundamental relation (10) which holds P^x-almost surely for all $x \in B$. Conditions (a) and (b) are trivial while condition (c) reduces precisely to Lemma 2. Now Theorem 3 can be deduced from Theorem 2 by applying it to the double multiplicative functional

$$\tilde{M}_t M_t = 1_{\{t<\tau_B\}} M_t, \qquad t > 0,$$

where M_t is as in Theorem 2.

References

[1] K. L. Chung, Lectures from Markov Processes to Brownian Motion, Grundlehren der mathematischen Wissenschaften 249, Springer-Verlag 1982.

[2] K. L. Chung and R. K. Getoor, The condenser problem, Ann. Probab. 5 (1977), 82-86.

[3] P. Courrège and P. Priouret, Axiomatique du problème de Dirichlet et processus de Markov, Sém. Brelot-Choquet-Deny (Théorie de potentiel), 8e année, 1963/64.

[4] I. V. Girsanov, Strongly-Feller processes, I. General properties, Theory Probab. Appl. (translated from the Russian) 5 (1960), 5-24.

[5] R. Z. Khas'minskii, On positive solutions of the equation Au + Vu = 0, Theory Probab. Appl. (translated from the Russian) 4 (1959), 309-318.

K. L. Chung
Department of Mathematics
Stanford University
Stanford, California 94305

Seminar on Stochastic Processes, 1985
Birkhäuser, Boston, 1986

ANOTHER LOOK AT WILLIAMS' DECOMPOSTION THEOREM

by

P. J. Fitzsimmons

1. Introduction

In studying the excursions of a diffusion process above its past minimum level, we have discovered a conceptually simple proof of Williams' decomposition [5] of a transient diffusion at its global minimum. We use an approximation argument based on the trivial observation that the minimum level of the diffusion is the smallest y such that $T_y < +\infty$, $T_{y-} = +\infty$, where T_y is the hitting time of y.

2. Statement of the Theorem

Let $X = (\Omega, \mathscr{F}, \mathscr{F}_t, \theta_t, X_t, P^x)$ be a regular diffusion on $]A,B[\subset \mathbb{R}$, with no killing on $]A,B[$. We take Ω to be the space of paths $\omega : [0,+\infty[\to]A,B[\cup \{\Delta\}$ which are absorbed in Δ at time $\zeta(\omega)$ and which are continuous on $[0,\zeta(\omega)[$. The killing operators k_t are defined as usual and $\mathscr{F}^0 = \sigma\{X_t, t > 0\}$, where X_t is the coordinate map $\omega \to \omega(t)$.

Let s be a scale function for X. To ensure a global minimum for X we assume that $-\infty = s(A) < s(B) < +\infty$ and that $\lim_{t\uparrow\zeta} X_t = B$. Consequently, setting $T_y = \inf(t > 0 : X_t = y)$, we must have $T_y \uparrow \zeta$ as $y \uparrow B$. Thus

$$(2.1) \quad P^x(T_y < +\infty) = h(x)/h(y), \qquad A < y < x < B,$$

where $h \equiv s(B) - s$ (see [1]).

Now fix $b \in]A,B[$ and for $x \in [A,b[$ let $P_x^{\downarrow}$ denote the law on $(\Omega, \mathscr{F}^0)$ under which the coordinate process behaves like X started at b, conditioned to converge to A, and then killed at time T_x. More precisely, for $F \in b\mathscr{F}^0$,

$$(2.2) \qquad P_x^{\downarrow}(F) = P^{b/h}(F \circ k_{T(x)}) \qquad (T(x) \equiv T_x);$$

here $P^{b/h}$ is the law of the h-transform of X with $h = s(B) - s$ as above. $P_x^{\downarrow}$ is the law of a diffusion with initial distribution ε_b and semigroup

$$P_t^{\downarrow}f(y) = P^y((fh)(X_t); \, t < T_x)/h(y), \; x < y < b.$$

Secondly, let $P_x^{\uparrow}$ denote the law of X started at x and conditioned to converge to B without returning to x after time $t = 0$. The transition semigroup corresponding to $P_x^{\uparrow}$ is

$$P_t^{\uparrow}f(y) = P^y((fg)(X_t))/g(y), \; x < y < B,$$

where $g = g_x = 1 - h/h(x)$.

Finally, let $\gamma = \inf(X_t : 0 \leq t < \zeta)$ and $\rho = \inf(t : X_t = \gamma)$. Williams' decomposition [5, T2.4] can now be stated as

(2.3) THEOREM.

a) $P^b(\gamma > x) = 1 - h(b)/h(x)$, $x \leq b$;

b) ρ _is the unique time_ t _at which_ $X_t = \gamma$, _almost surely_ P^b;

c) _for_ $F, G \in b\mathscr{F}^0$ _and_ $\phi \in b\mathscr{B}(\mathbb{R})$,

(2.4) $$P^b(F \circ k_\rho \phi(\gamma) G \circ \theta_\rho) = P^b(P^{\downarrow}_{\gamma}(F)\phi(\gamma)P^{\uparrow}_{\gamma}(G)).$$

In other words, $(X_t, 0 \leq t < \rho)$ and $(X_{\rho+t}, t \geq 0)$ are conditionally independent given γ, and given that $\gamma = y$, the law of the first process is $P^{\downarrow}_y$ while that of the second is $P^{\uparrow}_y$.

3. Proof of the theorem

Point a) is clear from (2.1) since $\{\gamma > x\} = \{T_x = +\infty\}$. We omit a proof of point b) but see [5] for an argument of the same point in the case of the Brownian bridge.

Before proceeding with the details, we outline the idea behind our proof of (2.4). Subdivide $]A,b[$ into intervals, all but possibly the lowest being of width $1/m$, where $m \in \mathbb{N}$. Of these intervals there is a unique one, say $[b - (j + 1)/m, b - j/m]$, which is hit by X but below which X does not travel. Let $\rho(m)$ (resp. $\sigma(m)$) be the hitting time (resp. exit time) of this interval. Then $(X_t,$

$0 < t < \rho(m))$ has law $P^{\downarrow}_{b-j/m}$ while $(X_{\sigma(m)+t}, t > 0)$ has law $P^{\uparrow}_{b-j/m}$. Moreover, because of (2.3b), $\rho(m) \uparrow \rho$ and $\sigma(m) \downarrow \rho$ as $m \to +\infty$; the path fragment $(X_t, \rho(m) < t < \sigma(m))$ thus evaporates in the limit as $m \to +\infty$.

Proceeding with the details, first observe that it suffices to consider F and G in (2.4) of the form $\Pi_{i=1}^{n} f_i(X(t_i))$ where $0 < t_1 < t_2 < \ldots < t_n$ and each f_i is continuous with support in a compact subset of $]A,B[$. **In the sequel we always assume that F,G take the above form.** Now for $x \in]A,b[$ and $t > 0$, $P^b(T_x = t) = 0$ and $P^b(T_{x+} = T_x = T_{x-}) = 1$. Thus, since $F \circ k_{T(x)} = F \cdot 1_{\{t(n)<T(x)\}}$, we have $P^b(\lim_{y\to x} F \circ k_{T(y)} = F \circ k_{T(x)}) = 1$ for each $x \in]A,b[$. Also, $P^b(T_y < +\infty, y > x; T_x = +\infty) = 0$ $(x \in A,b[)$; thus $P^b(\lim_{y\to x} 1_{\{T(y)<+\infty\}} = 1_{\{T(x)<+\infty\}}) = 1$. These observations combined with (3.1) below yield the fact that $x \to P^{\downarrow}_x(F)$ is continuous on $]A,b[$;

$$(3.1) \qquad P^{\downarrow}_x(F) = P^{b/h}(F \circ k_{T(x)})$$

$$= P^b(F \circ k_{T(x)} h(X_{T(x)}))/h(b)$$

$$= P^b(F \circ k_{T(x)}; T(x) < +\infty)h(x)/h(b).$$

Likewise, $x \to P^{\uparrow}_x(G)$ is continuous on $]A,b[$. To see this, let $L_x = \sup(t : X_t = x)$. It is easy to check that for each $c \in]A,b[$ and $x \in [c,b[$, $P^c(\lim_{y\to x} G \circ \theta_{L(y)} = G \circ \theta_{L(x)}) = 1$ (note that $P^c(0 < L(x) < \zeta)$ since $X_{\zeta-} = B$). But from the last exit results of [3] or [2] we know that

$$(3.2) \qquad P^{\uparrow}_x(G) = P^c(G \circ \theta_{L(x)}), \ A < c < x < b;$$

and the claimed continuity follows.

Now for $m \in \mathbb{N}$, $j \in \mathbb{N}$ define

$$S(j) = S(j,m) = T_{b-j/m},$$

$$K(m) = \sup(j \geq 1 : S(j,m) < +\infty),$$

$$\rho(m) = S(K(m),m), \ \gamma(m) = b - K(m)/m,$$

$$\sigma(m) = L_{\gamma(m)} = \rho(m) + L_{\gamma(m)}(\theta_{\rho(m)}).$$

As $m \to +\infty$, $\gamma(m) \to \gamma$; by (2.3b), $\rho(m) \uparrow \rho$ and $\sigma(m) \downarrow \rho$. Also, $F \circ k_{\rho(m)} = F \cdot 1_{\{t_n < \rho(m)\}} \to F \circ k_\rho$ since $1_{\{t_n < \rho(m)\}} \uparrow 1_{\{t_n < \rho\}}$; similarly $G \circ \theta_{\sigma(m)} \to G \circ \theta_\rho$ as $m \to +\infty$. Thus if ϕ is continuous with support in a compact subset of $]A,b[$,

(3.3) $$P^b(F \circ k_\rho \phi(\gamma) G \circ \theta_\rho)$$

$$= \lim_{m\to+\infty} P^b(F \circ k_{\rho(m)} \phi(\gamma(m)) G \circ \theta_{\sigma(m)}),$$

by bounded convergence. Let us now evaluate the right hand side of (3.3), using the strong Markov property at each $S(j)$, together with (3.1) and (3.2):

(3.4) $$P^b(F \circ k_{\rho(m)} \phi(\gamma(m)) G \circ \theta_{\sigma(m)})$$

$$= \sum_{j\geq 1} P^b(F \circ k_{S(j)};\ S(j) < \infty)$$

$$\times\ \phi(b - j/m) P^{b-j/m}(G \circ \theta_{L(b-j/m)};\ S(j+1) = +\infty)$$

$$= \sum_{j\geq 1} P^b(F\ k_{S(j)} | S(j) < +\infty) P^{b-j/m}(G\ \theta_{L(b-j/m)} | S(j+1) = +\infty)$$

$$\times\ P^b(S(j) < +\infty,\ S(j+1) = +\infty)\phi(b - j/m)$$

$$= \sum_{j\geq 1} P^{\downarrow}_{b-j/m}(F) P^{\uparrow}_{b-j/m}(G) P^b(\gamma(m) = b - j/m)\phi(b - j/m)$$

$$= P^b(P^{\downarrow}_{\gamma(m)}(F)\phi(\gamma(m))P^{\uparrow}_{\gamma(m)}(G)).$$

Using the continuity properties of $x \to P^{\downarrow}_x(F)$ and $x \to P^{\uparrow}_x(G)$ asserted earlier, we may now let $m \to +\infty$ in (3.4) and (3.3) to obtain (2.4).

(3.5) REMARK. One can show that $P^{\uparrow}_x = n^{\uparrow}_x(\cdot | T_{x+} = +\infty)$ where $n^{\uparrow}_x$ is the Itô excursion law of X above the level x. A more detailed discussion of the excursions of a diffusion above its past minimum will be undertaken in a future paper.

References

1. K. Ito and H. P. McKean. Diffusion Processes and their Sample Paths. Springer-Verlag, Berlin, 1965.

2. B. Maisonneuve. Exit Systems. Ann. Prob., 3(1975), 399-411.

3. P. A. Meyer, R. T. Smythe and J. B. Walsh. Birth and death of Markov processes. Proc. 6th Berk. Symp. Stat. Prob., Vol. III, 295-305. Univ. of Cal. Press, 1972.

4. D. Williams. Path decomposition and continuity of local time for one-dimensional diffusions, I. Proc. London Math. Soc. (3) 28 (1974), 738-768.

5. W. Vervaat. A relation between Brownian bridge and Brownian excursion. Ann. Prob., 7 (1979), 143-149.

P. J. Fitzsimmons
Department of Mathematical Sciences
The University of Akron
Akron, Ohio 44325

Seminar on Stochastic Processes, 1985
Birkhäuser, Boston, 1986

SOME REMARKS ON A THEOREM OF DYNKIN

by

R. K. Getoor*

In [1], Dynkin gave a very simple proof of Nagasawa's theorem (see [3] or [4]) and its generalization [2]. His proof is based on a simple general result on measurable flows (see [1], Theorems 1 and 2). Unfortunately in the application to Nagasawa's theorem in section 3, there is a measurability difficulty. This is hinted at in the footnote on page 616 of [1]. The purpose of this note is to correct this difficulty and to prove a theorem that shows that there really is a measurability problem that must be addressed. We use the notation of [1] without special mention.

Let $(E, \mathcal{E})$ be a standard Borel space and suppose that $p = (p_t)$ and $\hat{p} = (\hat{p}_t)$ are stationary transition functions on $(E, \mathcal{E})$ that are in weak duality relative to a σ-finite measure m. Let $\mu = (\mu_t)_{t>0}$ and $\nu = (\nu_t)_{t>0}$ be entrance rules for p and $\hat{p}$ respectively. Suppose that $\overline{\mu} = \int_R \mu_t dt$

*Research support, in part, by NSF Grant DMS 8919377.

and $\bar{\nu} = \int_R \nu_t dt$ are σ-finite and absolutely continuous with respect to m. According to (3.6) of [2] one may suppose that $\bar{\mu} = u \cdot m$ and $\bar{\nu} = v \cdot m$ where u and v are co-excessive and excessive respectively, and further that u and v are Borel and finite a.e. m. We claim that $\mu_t(v = \infty) = 0$ and $\nu_t(u = \infty) = 0$ for each t. To see this let $A = \{v < \infty\}$ and $B = \{v = \infty\}$. Since v is excessive, A is absorbing so that $p_t(x,B) = 0$ for $t > 0$ when $x \in A$. If $\mu_s(B) = 0$, then $\mu_s P_t(B) = \int_A \mu_s(dx) P_t(x,B) = 0$ for all $t > 0$. Given $t \in \mathbb{R}$, since $\mu_s(B) = 0$ a.e. in s we may choose $s_n \uparrow t$ with $\mu_{s_n}(B) = 0$ for each n. Hence $0 = \mu_{s_n} P_{t-s_n}(B) \uparrow \mu_t(B)$ as $n \to \infty$. Similarly $\nu_t(u = \infty) = 0$ for each t. Let $\mathscr{E}_0$ be a countable family of positive $\mathscr{E}$ measurable functions which generates $\mathscr{E}$. Let Ω_r (resp. Ω_ℓ) be the space of all trajectories with random birth and death times α and β such that $t \to f \circ X_t(\omega)$ is right (resp. left) continuous on $]\alpha(\omega), \beta(\omega)[$ for each $f \in \mathscr{E}_0$. Here $X_t(\omega) = \omega(t)$ is the coordinate process. We assume that the measure P_μ^v may be constructed on $(\Omega_r, \mathscr{F}^{0,r})$ and that $\hat{P}_\nu^u$ may be constructed on $(\Omega_\ell, \mathscr{F}^{0,\ell})$. Here $\mathscr{F}^{0,r}$ (resp. $\mathscr{F}^{0,\ell}$) is the σ-algebra generated by the coordinate maps $\{X_t : t \in \mathbb{R}\}$ restricted to Ω_r (resp. Ω_ℓ). If $\theta_t \omega(s) = \omega(s + t)$ for $s,t \in \mathbb{R}$ then $\theta_t : \Omega_r \to \Omega_r$ and $\theta_t : \Omega_\ell \to \Omega_\ell$.

Define a "backward" transition function by $q_{-t}(dy,x) = \hat{p}_t(x,dy)$ for $t > 0$ and $\nu^*_{-t} = \nu_t$ for all $t \in \mathbb{R}$. Then one easily checks that if $t_1 < \dots < t_n$,

(1.1) $\hat{P}_\nu^u(\alpha < t_1, X_{t_1} \in dx_1, \dots, X_{t_n} \in dx_n, t_n < \beta)$

$$= Q_u^{\nu^*}(\alpha < -t_n, X_{-t_n} \in dx_n, \ldots, X_{-t_1} \in dx_1, -t_1 < \beta),$$

where $Q_u^{\nu^*}$ is the measure defined in (3.4) of [1]. Define the reversal operator $R : \Omega_r \cup \Omega_\ell \to \Omega_r \cup \Omega_\ell$ by $R\omega(t) = \omega(-t)$. Note that $R : \Omega_r \to \Omega_\ell$ (resp. $R : \Omega_\ell \to \Omega_r$) and that R is $\mathcal{F}^{0,r}|\mathcal{F}^{0,\ell})$ (resp. $\mathcal{F}^{0,\ell}|\mathcal{F}^{0,r})$ measurable on Ω_r (resp. Ω_ℓ). It follows from (1.1) that

$$(1.2) \qquad Q_u^{\nu^*} = R\hat{P}_\nu^u; \qquad RQ_u^{\nu^*} = \hat{P}_\nu^u;$$

the first relation holding on Ω_r (since $\hat{P}_\nu^u$ is on Ω_ℓ) and the second on Ω_ℓ. Note the crucial fact for applying Theorems 1, 2, and 3 of [1] that $(t,\omega) \to X_t(\omega)$ is $\mathcal{B} \times \mathcal{F}^{0,r}$ (resp. $\mathcal{B} \times \mathcal{F}^{0,\ell}$) measurable on Ω_r (resp. Ω_ℓ). Here $\mathcal{B}$ is the Borel σ-algebra on $\mathbb{R}$. Therefore the flows θ_t on Ω_r and Ω_ℓ are measurable as defined in Section 1.1 of [1].

Suppose τ is a stationary time on Ω_ℓ (resp. Ω_r). See subsection 1.1 of [1]. Define

$$(1.3) \qquad \tau^*(\omega) = -\tau(R\omega) \quad \text{on} \quad \Omega_r \quad (\text{resp. } \Omega_\ell).$$

It is readily checked that τ^* is a stationary time on Ω_r (resp. Ω_ℓ), and that $\theta_\tau R = R\theta_{\tau^*}$ in either case. Now suppose that τ is stationary on Ω_r. Then Theorem 3 of [1] is applicable and gives

$$(1.4) \qquad \theta_\tau Q_u^{\nu^*} = \theta_\tau P_\mu^\nu$$

as measures on Ω_r. Combining this with (1.2) we have

$$\theta_\tau R\hat{P}^u_\nu = \theta_\tau Q^{\nu^*}_u = \theta_\tau P^v_\mu.$$

But $\theta_\tau R\hat{P}^u_\nu = R\theta_{\tau^*}\hat{P}^u_\nu$ and since $X_t(R\theta_{\tau^*}) = X_{\tau^*-t}$ on Ω_ℓ we see that

$$(1.5) \qquad (X_{\tau^*-t}, \hat{P}^u_\nu) \cong (X_{\tau+t}, P^v_\mu)$$

where "$\cong$" means that the two processes have the same law. Because $\hat{P}^u_\nu$ is on Ω_ℓ while P^v_u is on Ω_r, $t \to f \circ X_{\tau^*-t}$ (resp. $t \to f \circ X_{\tau+t}$) is right continuous almost surely under $\hat{P}^u_\nu$ (resp. P^v_μ) for $f \in \mathscr{E}_0$. Finally suppose $\tau = \alpha$, then $\tau^* = \beta$. If μ is an entrance **law** at 0 for p, then $\alpha = 0$ a.s. P^v_μ and we obtain

$$(1.6) \qquad (X_{\beta-t}, \hat{P}^u_\nu) \cong (X_t, P^v_\mu).$$

The formulas (1.5) and (1.6) are correct versions of Theorems 4 and 5 of [1].

For simplicity suppose now that E is identified with a Borel subset of a compact metric space and $\mathscr{E}$ is the Borel σ-algebra. Choosing $\mathscr{E}_0$ to be a countable collection of uniformly continuous functions that is uniformly dense in the bounded uniformly continuous functions on E, Ω_r (resp. Ω_ℓ) becomes the space of all right (resp. left) continuous trajectories. This is the case of most interest for Markov processes. If $\hat{P}^u_\nu$ may be constructed on Ω_r (as well as on Ω_ℓ) so that $X^- = (X_{t-})$ exists a.s. $\hat{P}^u_\nu$ on Ω_r and $(X^-, \Omega_r, \hat{P}^u_\nu)$ is equivalent to $(X, \Omega_\ell, \hat{P}^u_\nu)$, then from (1.6),

$(X^{-}_{\beta-t}, \hat{P}^{u}_{\nu}) \cong (X_t, P^{v}_{\mu})$. One easily checks that $X^{-}_{\beta-t} = X_{(\beta-t)+}$ and this gives the usual form of Nagasawa's theorem and its extension [2].

Similarly starting with $RQ^{\nu*}_{u} = P^{u}_{\nu}$ and a stationary time τ on Ω_ℓ on obtains

(1.7) $$(X_{\tau+t}, \hat{P}^{u}_{\nu}) \cong (X_{\tau^*-t}, P^{v}_{\mu}),$$

and taking $\tau = \alpha$ and ν and entrance **law** at 0

(1.8) $$(X_t, \hat{P}^{u}_{\nu}) \cong (X_{\beta-t}, P^{v}_{\mu}), \quad t > 0.$$

Under the set-up of the previous paragraph (i.e., Ω_r and Ω_ℓ are the right and left continuous trajectories respectively) both X_t and $X_{\beta-t}$ are left continuous almost surely under $\hat{P}^{u}_{\nu}$ and P^{v}_{μ} respectively. If X_{t+} (resp. X_{t-}) exists under $\hat{P}^{u}_{\nu}$ (resp. P^{v}_{μ}) one obtains from (1.8)

(1.9) $$(X_{t+}, \hat{P}^{u}_{\nu}) \cong (X_{(\beta-t)+}, P^{v}_{\mu}), \quad t > 0.$$

In particular if (X_{t+}) is equivalent to (X_t) under $\hat{P}^{u}_{\nu}$ this again is the usual form of the reversal theorems.

We next show that if Ω is a set of trajectories that contains enough discontinuous functions, then $(t,\omega) \to X_t(\omega) = \omega(t)$ is **not** $\mathscr{B} \times \mathscr{F}^0$ measurable where $\mathscr{F}^0$ is the σ-algebra generated by the coordinate maps. To be precise ω denotes a real valued path defined on a non-void open interval $]\alpha(\omega),\beta(\omega)[$. Let Ω_1 (resp. Ω_2) be the set of paths that are right continuous with left limits (resp. left continuous with right limits) on $]\alpha,\beta[$. Let ϕ be defined on $\Omega_1 \cup \Omega_2$ by $\phi\omega(t) = \omega(t-)$ if $\omega \in \Omega_1$, $\phi\omega(t) = \omega(t+)$ if $\omega \in \Omega_2$. Then ϕ maps Ω_1 onto Ω_2 and Ω_2 onto Ω_1. Also ϕ is the identity on $\Omega_1 \cap \Omega_2$. Let $D(\omega)$ denote the set of discontinuities of a path ω.

(1.10) PROPOSITION. Let $\Omega' \subset \Omega_1$ have the property that if J is a countable subset of R there exists $\omega_J \in \Omega'$ with $D(\omega_J) \neq \phi$ and $D(\omega_j) \cap J = \phi$. (Here ϕ is the empty set.) Let $\Omega'' = \phi(\Omega') \subset \Omega_2$. Let Ω be a set of trajectories with $\Omega' \cup \Omega'' \subset \Omega$. Then $(t,\omega) \to X_t(\omega) = \omega(t)$ is not $\mathscr{B} \times \mathscr{F}^0$ measurable where $\mathscr{F}^0 = \sigma(X_t : t \in \mathbb{R})$.

REMARKS. Of course, $\Omega' = \Omega_1$ satisfies the hypotheses of (1.10). For many processes Ω_1 (or Ω_r the right continuous trajectories) is the natural sample space. However, if we want the sample space closed under the reversal operator, R, one would have to enlarge it to contain $\Omega_1 \cup \Omega_2$, and this destroys any hope of measurability according to (1.10).

PROOF. The σ-algebra $\mathscr{B} \times \mathscr{F}^0$ is generated by functions of the form

$$(1.11) \qquad F(t,\omega) = g(t)f_1(X_{s_1}(\omega)) \ldots f_m(X_{s_m}(\omega))$$

where $g \in \mathscr{B}$, $f_1,\ldots,f_m \in \mathscr{E}$, and $s_1 < \ldots < s_m$. (We set $f(X_s) = 0$ if s is not in the interval $]\alpha,\beta[$.) Define $\phi : \Omega \to \Omega$ by $\phi\omega = \omega$ if $\omega \in \Omega - (\Omega' \cup \Omega'')$ and ϕ as above on $\Omega' \cup \Omega''$. Then $\phi(\Omega') = \Omega''$ by assumption and so $\phi(\Omega'') = \Omega'$. Therefore $\phi : \Omega \to \Omega$. If F is of the form (1.11) then

$$\{\omega : F(\cdot,\omega) \neq F(\cdot,\phi\omega)\} \subset \{\omega \in \Omega' \cup \Omega'' : D(\omega) \cap \{s_1,\ldots,s_m\} \neq \phi\}.$$

It then follows from the monotone class theorem if $F \in \mathscr{B} \times \mathscr{F}^0$, there exists a countable set $J(F) \subset \mathbb{R}$ that is independent of ω such that

$$\{\omega : F(\cdot,\omega) \neq F(\cdot,\phi\omega)\} \subset \{\omega \quad \Omega' \cup \Omega'' : D(\omega) \cap J(F) \neq \phi\}.$$

Now suppose that $X(t,\omega) = X_t(\omega)$ is $\mathscr{B} \times \mathscr{F}^0$ measurable. Let $i(x) = x$ on $\mathbb{R}$. Then $i \circ X(t,\omega) \in \mathscr{B} \times \mathscr{F}^0$ and so there exists a countable set $J \subset \mathbb{R}$ such that

$$\{\omega : i \circ X(\cdot,\omega) \neq i \circ X(\cdot, \phi\omega)\} \subset \{\omega \in \Omega' \cup \Omega'' : D(\omega) \cap J \neq \phi\}.$$

Let ω_J be as in the hypotheses of (1.10). Then $X(\cdot,\omega_J) = X(\cdot,\phi\omega_J)$ on $]\alpha(\omega_J),\beta(\omega_J)[$ which implies that $D(\omega_J) = \phi$, completing the proof of (1.10).

We conclude with a simple example that shows that there is a real difficulty in Theorem 4 of [1] as stated. Let E be the union of two disjoint intervals which we write as $]-1,0^-] \cup [0^+,1[$ with the implied order. The process X is translation to the right at unit speed killed on leaving E except that as it approaches 0^- it jumps to 0^+ and starting at 0^- it remains there for an exponential time and then dies. $\hat{X}$ is defined similarly except in the opposite direction. One may easily write down the transition functions p and $\hat{p}$, but we shall not record them explicitly. If m is Lebesgue measure on E, then p and $\hat{p}$ are in weak duality and the hypotheses of Theorem 4 of [1] are satisfied. Let $\mu = (\mu_t)$ be the entrance law for p corresponding to starting at -1 and $\nu = (\nu_t)$ the entrance law for $\hat{p}$ corresponding to starting at 1. Then $\bar{\mu} = \bar{\nu} = m$ and $u = v = 1$. Note that almost surely P_μ, $\beta = 2$ so that $X_{\beta-1} = 0^+$, while almost surely $\hat{P}_\nu$, $X_1 = 0^-$. Thus (1.8) fails for this example.

REMARK. Dynkin has pointed out that another approach is to use the space $\Omega_r \times \Omega_\ell$ in place of $\Omega_r \cup \Omega_\ell$. This amounts to considering the process in "split" time.

References

1. E. B. Dynkin. An application of flows to time shift and time reversal in stochastic processes. Trans. Amer. Math. Soc. 287 (1985), 613-619.

2. R. K. Getoor and J. Glover. Riesz decompositions in Markov process theory. Trans. Amer. Math. Soc. 285 (1984), 107-132.

3. R. K. Getoor and M. J. Sharpe. Naturality, standardness, and weak duality for Markov processes. Z. Wahrscheinlichkeitstheorie verw. Geb. 67 (1984), 1-62.

4. M. J. Sharpe. Some transformations of diffusions by time reversal. Ann. Prob. 8 (1980), 1157-1162.

R. K. Getoor
Department of Mathematics
University of California, San Diego
La Jolla, California 92093

Seminar on Stochastic Processes, 1985
Birkhäuser, Boston, 1986

SOME REMARKS ON MEASURES ASSOCIATED WITH HOMOGENEOUS RANDOM MEASURES

by

R. K. Getoor*

1. Introduction

This paper is an extension of some of the results in [3] and [4]. As in §8 of [4] we assume that X is a Borel right process with Lusin state space $(E, \mathcal{E})$ such that $X_{t-}(\omega)$ exists in E for $0 < t < \zeta(\omega)$ and $\omega \in \Omega$. (Some of our results do not depend on the existence of left limits as will be clear from the context. However, for simplicity of exposition we shall assume this hypothesis throughout this paper.) In addition, we fix an arbitrary σ-finite excessive measure, m.

Section 2 contains some weak limit theorems for Revuz measures. These results are immediate consequences of Theorem 8.7 of [4]. The reason for insisting on them here, in addition to their intrinsic interest, is that they give

---

*Research supported in part by NSF Grant MCS 79-23922.

an immediate and elementary proof of a recent result of Cheval and Feldman [1]. Cheval and Feldman obtain a weak limit theorem for the capacitary measure of a relatively compact domain in a (complete) Riemannian manifold with Ricci curvature bounded from below. Their proof uses estimates of Cheeger-Yau type and the asymptotic properties of Legendre functions $P_{\nu}^{1-n/2}$ as $\nu \to \infty$. A number of other interesting results are contained in [1].

In sections 3 and 4 we extend some of the results of [3] and [4] in two directions. In the first place, we consider Revuz measures in general rather than just capacitary measures, and secondly, we make no duality assumptions. As in section 2 the arguments are elementary in the extreme.

2. A Weak Limit Theorem

Theorem 8.7 of [4] states that if κ is an integrable homogeneous random measure (HRM) and f is a bounded, positive, $\mathscr{E}^*$ measurable function such that $t \to f(X_{t-})$ is P^m a.s. left continuous on $]0,\zeta[$, then

$$(2.1)\quad \lim_{t\downarrow 0} t^{-1}E^m \int_0^t f(X_{s-})\kappa(ds) = \lim_{t\downarrow 0} t^{-1}E^m(f(X_0)\kappa]0,t]).$$

In (2.1) the limit on the left exists for any positive $\mathscr{E}^*$ measurable f, and is, by definition, $\nu_\kappa(f)$ where $\nu_\kappa = \nu_\kappa^m$ is the Revuz measure of κ relative to m. In order to save parentheses we write $\kappa]0,t]$ in place of $\kappa(]0,t])$ provided no confusion is possible. The proof of (2.1) uses Fubini's theorem and elementary manipulations. Applying (2.1) to bounded continuous functions one obtains:

(2.2) COROLLARY. *Let* κ *be an integrable* HRM. *Then*

$$t^{-1}E^{x}\kappa]0,t]m(dx) \overset{b}{\to} \nu_{\kappa}(dx) \quad \textit{as} \quad t \downarrow 0.$$

Here "$\overset{b}{\to}$" stands for Bernoulli (or weak) convergence, that is, $\mu_t \overset{b}{\to} \nu$ means that $\int f\, d\mu_t \to \int f\, d\nu$ for all bounded continuous functions on E. Corollary 2.2 follows from (2.1) because

$$E^{m}(f(X_0)\kappa]0,t]) = \int f(x)E^{x}\kappa]0,t]m(dx).$$

Before giving some applications of (2,2) we point out that the following result may be proved by almost the same argument as that used in the proof of Theorem 8.7 in [4]. It is a "right hand" analog of (2.1) and does not require the existence of left limits.

(2.3) THEOREM. *Let* κ *be an integrable* HRM *and* f *a bounded, positive,* $\mathscr{E}^*$ *measurable function such that* $t \to f(X_t)$ *is* P^m a.s. *right continuous, then*

$$\text{(2.4)} \quad \lim_{t\downarrow 0} t^{-1}E^{m}(f(X_t)\kappa]0,t]) = \lim_{t\downarrow 0} t^{-1}E^{m}\int_0^t f(X_s)\kappa(ds).$$

In section 4 we shall give some extensions of (2.4) to non-integrable κ.

Let U_κ be the **potential kernel** of κ, that is

$$\text{(2.5)} \qquad U_{\kappa}f(x) = E^{x}\int_0^{\infty} f(X_{t-})\kappa(dt)$$

for $f \in \mathcal{E}^{*+}$. If $m = \mu U$, it follows from (8.5) of [4], that $\nu_\kappa = \nu_\kappa^m = \mu U_\kappa$. Combining this observation with (2.2) we see that

$$(2.6) \qquad t^{-1} E^x \kappa]0,t] \mu U(dx) \overset{b}{\to} \mu U_\kappa(dx)$$

as $t \to 0$ for any HRM, κ, such that $\mu U_\kappa(1) < \infty$. This is the condition that κ be integrable relative to $m = \mu U$. In particular, if $\mu = \varepsilon_y$ and $U_\kappa(y,E) < \infty$, one has as $t \downarrow 0$

$$(2.7) \qquad t^{-1} E^x \kappa]0,t] U(y,dx) \overset{b}{\to} U_\kappa(y,dx).$$

This expresses the kernel U_κ in terms of the potential kernel U of X and the **characteristic** $c_t^\kappa(x) = c_t(x) = E^x \kappa]0,t]$ of κ.

Suppose now that $B \in \mathcal{E}$ is **strongly transient** in the sense that $P^m(L_B > \zeta) = 0$ where $L_B = \sup \{t : X_t \in B\}$ and the supremum of the empty set is taken to be zero. Then $\kappa(dt) = 1_{\{L_B > 0\}} \varepsilon_{L_B}(dt)$ is a HRM and its Revuz measure $\pi_B = \nu_\kappa^m$ is called the capacitary measure of B (relative to m). It follows from [2] and [3], that much weaker conditions than the strong transience of B suffice for what follows, but to keep technicalities and definitions to a minimum we shall assume B strongly transient. The **capacity** of B, C(B), is defined by $C(B) = \pi_B(1)$, the total mass of π_B. Sufficient conditions for $C(B) < \infty$ are known under various assumptions. See [3] for example. Applying (2.2) to this situation yields the following result.

(2.8) PROPOSITION. Suppose B is strongly transient and $C(B) < \infty$. Then as $t \to 0$,

$$t^{-1}P^x[0 < L_B < t]m(dx) \xrightarrow{b} \pi_B(dx).$$

Let $T_B = \inf\{t > 0 : X_t \in B\}$ be the hitting time of B where, as usual, the infimum of the empty set is plus infinity. It is well known, see (6.20) of [4], for example that $P^m(T_B = t) = 0$ for each fixed $t > 0$. Similarly $P^m(L_B = t) = 0$ for $t > 0$. Observe that

$$\{0 < L_B < t\} \subset \{T_B < t, T_B \circ \theta_t = \infty\} \subset \{0 < L_B \le t\}.$$

Combining these observations the conclusion of (2.8) may be written

$$(2.9) \quad t^{-1}P^x[T_B < t, T_B \circ \theta_t = \infty]m(dx) \xrightarrow{b} \pi_B(dx)$$

as $t \to 0$, and this is precisely the result of Cheval and Feldman [1] mentioned in the introduction.

3. Some Asymptotic Results

By definition, the Revuz measure $\nu_\kappa = \nu_\kappa^m$ of a HRM relative to m is given by

$$(3.1) \qquad \nu_\kappa^m(f) = \lim_{t \downarrow 0} t^{-1}E^m \int_0^t f(X_{s-})\kappa(ds)$$

if $f \ge 0$ is $\mathcal{E}^*$ measurable. Now m may be written uniquely $m = m_i + m_p$ where m_i is **invariant** ($m_i P_t = m_i$ for $t > 0$) and

m_p is **purely excessive** (m_p is excessive and $m_pP_t(B) \to 0$ as $t \to \infty$ if $B \in \mathcal{E}$ with $m_p(B) < \infty$). See, for example, [2]. The following result treats these two extreme cases.

(3.2) PROPOSITION. Let $f \in \mathcal{E}^*$ be positive and $\nu_\kappa^m(f) < \infty$. (a) If m is invariant, then

$$(3.3) \qquad E^m \int_0^t f(X_{s-})\kappa(ds) = t\nu_\kappa^m(f).$$

(b) If m is purely excessive, then

$$(3.4) \qquad \lim_{t\to\infty} t^{-1}E^m \int_0^t f(X_{s-})\kappa(ds) = 0.$$

As an immediate corollary of (3.2) we have the following complement to the definition (3.1),

$$(3.5) \qquad \lim_{t\to\infty} t^{-1}E^m \int_0^t f(X_{s-})\kappa(ds) = \nu_\kappa^{m_i}(f)$$

for $f \in \mathcal{E}^*$ with $f > 0$ and $\nu_\kappa^m(f) < \infty$.

In proving (3.3) and (3.4), replacing κ by $(f_-{*}\kappa)(ds) = f(X_{s-})\kappa(ds)$, we may suppose $f = 1$ without loss of generality. Let $c_t(x) = E^x\kappa]0,t]$ and $\gamma(t) = m(c_t)$. Then $\gamma(t+s) = \gamma(t) + mP_t(c_s) = \gamma(t) + \gamma(s)$ if m is invariant. Consequently $\gamma(t) = t\,\gamma(1)$ and then $\gamma(1) = \nu_\kappa^m(1)$, proving (3.3). For general m,

$$\nu_\kappa^m(1) = \lim_{t\to 0} t^{-1}\gamma(t) = \sup_{t>0} t^{-1}\gamma(t).$$

Thus if $\nu_\kappa^m(1) < \infty$, one has $\gamma(s) < \infty$ and so if m is purely excessive

$$\gamma(t+s)-\gamma(t)=mP_t(c_s)\to 0$$

as $t\to\infty$ for each fixed $s>0$. The conclusion (3.4) is now immediate in view of the following elementary and well-known lemma. We give a proof for the convenience of the reader.

(3.6) LEMMA. _Let_ $\phi :]0,\infty[\ \to \mathbb{R}^+$ _be increasing and subadditive. If_ $\phi(t+s)-\phi(t)\to 0$ _as_ $t\to\infty$ _for some_ $s>0$, _then_ $t^{-1}\phi(t)\to 0$ _as_ $t\to\infty$.

PROOF. For simplicity take $s=1$. If $0<y<x$ write

$$\phi(x)=\int_x^{x+1}[\phi(x)-\phi(t)]dt+\int_y^x[\phi(t+1)-\phi(t)]dt+\int_y^{y+1}\phi(t)dt$$

$$=\eta(x)+\int_y^x \Delta(t)dt+a(y).$$

Now $\Delta(t)=\phi(t+1)-\phi(t)\to 0$ as $t\to\infty$ and $\eta(x)=\int_0^1(\phi(x)-\phi(t+x))dt\to 0$ as $x\to\infty$ since $0\le\phi(t+x)-\phi(x)\le\phi(x+1)-\phi(x)$ if $t\le 1$. Thus

$$\frac{\phi(x)}{x}=\frac{\eta(x)+a(y)}{x}+\frac{1}{x}\int_y^x\Delta(t)dt\ ,$$

and first choosing y large and then letting $x\to\infty$ we obtain (3.6).

Next suppose B is strongly transient and $\kappa(dt)=\varepsilon_{L_B}(dt)\,1_{\{L_B>0\}}$. Then if $C(B)<\infty$ and f bounded, (3.5) becomes

$$(3.7)\qquad \lim_{t\to\infty} t^{-1}E^m[f(X_{L_B^-});\ 0 < L_B < t] = \nu_\kappa^{m_i}(f).$$

Recalling $\pi_B = \nu_\kappa^m$, one has by definition

$$(3.8)\qquad \lim_{t\downarrow 0} t^{-1}E^m[f(X_{L_B^-});\ 0 < L_B < t] = \pi_B(f).$$

Now assume that X and $\hat{X}$ are weak duality with respect to m. It is easy to see that if $\hat{\phi}(x) = \hat{P}^x(\zeta = \infty)$, then

$$(3.9)\qquad m_i(dx) = \hat{\phi}(x)m(dx).$$

Combining this with (2.1) gives $\nu_\kappa^{m_i}(f) = \nu_\kappa^m(\hat{\phi}f)$ because $\hat{\phi}$ is coexcessive and so $t \to \hat{\phi}(X_{t-})$ is P^m a.s. left continuous on $]0,\zeta[$. See (9.6) of [4]. Thus (3.5) and (3.7) become under the present hypotheses

$$(3.10)\qquad \lim_{t\to\infty} t^{-1}E^m\int_0^t f(X_{s-})\kappa(ds) = \nu_\kappa^m(f\hat{\phi})$$

$$(3.11)\qquad \lim_{t\to\infty} t^{-1}E^m[f(X_{L_B^-});\ 0 < L_B < t] = \pi_B(f\hat{\phi}).$$

Formula (3.11) is just (2.26) of [3] and so (3.5), (3.7), and (3.10) are extensions of this result.

4. Some Additional Remarks

If κ is a HRM and $f \geq 0$ is $\mathscr{E}^*$ measurable we define

$$(4.1)\qquad \mu_\kappa(f) = \mu_\kappa^m(f) = \sup_{t>0} t^{-1}E^m\int_{]0,t]} f(X_s)\kappa(ds)$$

$$= \lim_{t\downarrow 0} t^{-1} E^m \int_{]0,t]} f(X_s)\kappa(ds).$$

The equality of the supremum and the limit in (4.1) follows from the subadditivity of

(4.2) $$\phi(t) = E^m \int_{]0,t]} f(X_s)\kappa(ds).$$

In particular $2^n\phi(2^{-n})$ increases to $\mu_\kappa(f)$ as $n \to \infty$ and so μ_κ is a measure. Note the difference between μ_κ and ν_κ defined in (3.1). Also the limit on the right side of (2.4) is $\mu_\kappa(f)$. Define $f*\kappa(ds) = f(X_s)\kappa(ds)$ and $f_-*\kappa(ds) = f(X_{s-})\kappa(ds)$. Then $\mu_\kappa(f) = \nu_{f*\kappa}(1)$ and $\nu_\kappa(f) = \mu_{f_-*\kappa}(1)$. Both of these measures are special cases of the bimeasure of κ, β_κ, introduced by Sharpe in [7] but we shall not strive for the maximum generality and so shall not use the bimeasure. The next result is an extension of Theorem 2.3.

(4.3) THEOREM. <u>Let</u> κ <u>be a</u> HRM <u>such that</u> $\kappa = \Sigma\kappa^n$ <u>where each</u> κ^n <u>is an integrable</u> HRM <u>with</u> $t \to \kappa^n]0,t]$ <u>being adapted.</u> <u>If</u> f <u>is excessive, then</u>

(4.4) $$\lim_{t\downarrow 0} t^{-1} E^m(f(X_t)\kappa]0,t]) = \mu_\kappa(f).$$

PROOF. Note that

$$f(X_{t+s})\kappa]0,t+s] = f(X_{t+s})\kappa]0,t] + (f(X_s))\kappa]0,s]) \circ \theta_t,$$

and that

$$E^m(f(X_{t+s})\kappa]0,t]) = E^m(P_s f(X_t)\kappa]0,t]) \le E^m(f(X_t)\kappa]0,t])$$

because $\kappa]0,t]$ is $\mathscr{F}_t$ measurable and f is excessive. It follows that $\Psi(t) = E^m(f(X_t)\kappa]0,t])$ is subadditive. As a result the limit, call it L, on the left side of (4.4) exists and $2^k\Psi(2^{-k}) \uparrow L$ as $k \to \infty$. Let (f_p) be a sequence of bounded excessive functions with $f_p \uparrow f$ and let $\kappa_\ell = \sum_{n=1}^{\ell} \kappa^n$. Then $2^k E^m(f_p(X_{2^{-k}})\kappa_\ell]0,2^{-k}])$ increases with k, ℓ, and p. The left side of (4.4) is obtained by first letting p and ℓ approach infinity and then k, while in view of (2.3), $\mu_\kappa(f)$ is obtained by letting k, ℓ, and p approach infinity in that order. This establishes (4.3).

The next result seems innocuous but is quite useful when considering h-transforms.

(4.5) COROLLARY. Let κ be a HRM and $f \ge 0$ be $\mathscr{E}^*$ measurable. If κ satisfies the hypotheses of Theorem 4.3, then

$$\nu_\kappa(f) = \lim_{t\downarrow 0} t^{-1}E^m\left[\int_0^t f(X_{s-})\kappa(ds);\ t < \zeta\right].$$

PROOF. Replacing κ by $f_-*\kappa$ it suffices to suppose f = 1. Then using (4.4)

$$\lim_{t\downarrow 0} t^{-1}E^m(\kappa]0,t];\ t < \zeta)$$

$$= \lim_{t\downarrow 0} t^{-1}E^m(1_E(X_t)\kappa]0,t]) = \mu_\kappa(1) = \nu_\kappa(1).$$

(4.6) REMARK. The same argument shows that

$$\mu_\kappa(f) = \lim_{t\downarrow 0} t^{-1}E^m\left[\int_0^t f(X_s)\kappa(ds);\ t < \zeta\right].$$

Here is an application of (4.5). Let h be an excessive function with $m(h = \infty) = 0$. Let X^h be the h-transform of h. Since $P_t h = 0$ on $\{h = 0\}$, it is easy to see that hm is excessive for X^h. Here hm is the measure $h(x)m(dx)$. Let κ be a HRM satisfying the hypotheses of Theorem 4.3 and suppose, in addition, that for each n, $t \to \kappa^n]0,t]$ is finite (and hence right continuous) on $[0,\zeta[$. Then by a result of Meyer [6] one may suppose that each κ^n, and hence κ, is adapted to $(\mathcal{F}^*_{t+})$ and that the shift property $\kappa^n]0,t] \circ \theta_s = \kappa^n]s,t+s]$ holds identically in t, s, and ω. Hence each κ^n and κ may be regarded as a HRM of X^h. Let ν^h_κ and ν_κ be the Revuz measures of κ relative to (X^h, hm) and (X,m) respectively.

(4.7) THEOREM. <u>Let</u> κ, h, <u>and</u> m <u>be as above. Then</u> $\nu^h_\kappa = \nu_{h*\kappa}$.

PROOF. Let $\nu^h = \nu^h_\kappa$. Replacing κ by $f\_*\kappa$ where $0 \le f \le 1$, it suffices to show $\nu^h(1) = \nu_{h*\kappa}(1) = \mu_\kappa(h)$. Using Corollary 4.5 and the standard notation for h-transforms, see [5],

$$\nu^h(1) = \lim_{t\downarrow 0} t^{-1}E^{hm/h}(\kappa]0,t];\ t < \zeta).$$

But $\kappa]0,t]$ is $\mathcal{F}^*_{t+}$ measurable and so

$$E^{x/h}(\kappa]0,t];\ t < \zeta) = h(x)^{-1}E^x(h(X_t)\kappa]0,t])$$

if $0 < h(x) < \infty$. But hm does not change $\{h = 0\}$ and $m(h = \infty) = 0$ by hypothesis. Therefore

$$\nu^h(1) = \lim_{t\downarrow 0} \int_{\{h>0\}} m(dx)E^x(h(X_t)\kappa]0,t]).$$

Now $E^x h(X_t) = 0$ if $h(x) = 0$ and so using (4.3)

$$\nu^h(1) = \lim_{t\downarrow 0} t^{-1}E^m(h(X_t)\kappa]0,t]) = \mu_\kappa(h),$$

proving (4.7).

(4.8) REMARK. With the obvious notation the same argument shows that $\mu_\kappa^h = \mu_{h*\kappa} = h\mu_\kappa$.

(4.9) REMARK. The results in (4.5) and (4.6) may be extended to general HRM's as follows. Let κ be a HRM and $f > $ be $\mathcal{E}^*$ measurable. For (4.10) assume $f_-*\kappa$ is a HRM and for (4.11) assume $f*\kappa$ is a HRM. (Both conditions are satisfied if $0 < f < \infty$.) Then

$$\text{(4.10)}\qquad \nu_\kappa(f) = \lim_{t\downarrow 0} t^{-1}E^m\left[\int_0^t f(X_{s-})\kappa(ds);\ t < \zeta\right]$$

$$\text{(4.11)}\qquad \mu_\kappa(f) = \lim_{t\downarrow 0} t^{-1}E^m\left[\int_0^t f(X_s)\kappa(ds);\ t < \zeta\right].$$

The key point is to note that a simple modification of the proof of (8.10) in [4] yields

$$\liminf_{t\downarrow 0} t^{-1}E^m[g(X_0)\kappa]0,t];\ t < \zeta] \geq \nu_\kappa(g)$$

provided $g \geq 0$ and $t \to g(X_{t-})$ is left continuous a.s. P^m. Applying this (with $g = 1$) to the HRM's $f_-*\kappa$ and $f*\kappa$ one readily obtains (4.10) and (4.11).

References

1. I. Cheval and E. A. Feldman, The Wiener sausage, and a theorem of Spitzer in Riemannian manifolds. Preprint.

2. P. J. Fitzsimmons and B. Maisonneuve. Excessive measures and Markov processes with random birth and death. To appear Z. Wahrscheinlichkietstheorie verw. Geb.

3. R. K. Getoor. Capacity theory and weak duality. Seminar on Stochastic Processes, 1983, 97-130. Birkhauser, Boston, 1984.

4. R. K. Getoor and M. J. Sharpe. Naturality, standardness, and weak duality for Markov processes. Z. Wahrscheinlichkeitstheorie verw. Geb. 67 (1-62) 1984.

5. R. K. Getoor and J. Glover. Riesz decompositions in Markov process theory. Trans. Amer. Math. Soc. 285 (107-132) 1984.

6. P. A. Meyer. Ensembles aléatoires Markoviens homogènes I. Sem. Prob. VIII, 176-190. Springer Lectures Notes in Math. 381 (1974).

7. M. J. Sharpe. Discontinuous additive functionals of dual processes. Z. Wahrscheinlichkeitstheorie verw. Geb. 21 (81-95) 1972.

R. K. Getoor
Department of Mathematics
University of California, San Diego
La Jolla, California 92093

Seminar on Stochastic Processes, 1985
Birkhäuser, Boston, 1986

BROWNIAN EXIT DISTRIBUTION OF A BALL

by

Pei Hsu*

Summary

Let B be a ball in R^d and $X = \{X_t, t \geq 0\}$ be the standard Brownian motion in R^d. Define $\tau_B = \inf\{t > 0 : X_t \notin B\}$, the first exit time of X from the ball. We compute explicitly the transition density function of the killed Brownian motion $X^o = \{X_t, t < \tau_B\}$ and the joint distribution of $(\tau_B, X(\tau_B))$. A result of Wendel [5] is deduced as a simple consequence of the explicit joint density function.

Let D be a bounded domain in R^d and $X = \{X_t, t \geq 0\}$ the standard Brownian motion in R^d. The first exit time of X for domain D is defined to be

$$\tau_D = \inf\{t > 0 : X_t \notin D\}.$$

* Work supported in part by the grant NSF-MCS-82-01599

The first exit is $X(\tau_D)$. Because of the sample path continuity of Brownian motion, $X(\tau_D)$ lies on ∂D.

In this note we compute explicitly the joint distribution of (τ_D, X_{τ_D}) when $D = B$, the ball of radius 1 centered at the origin. The method used here was indicated in [1]. Previously, Wendel [5] computed the expectations of a family of functions of (τ_B, X_{τ_B}), which determines uniquely the joint distribution; but the explicit joint distribution was not given. As we will see later, Wendel's result can be obtained from our explicit density function by a simple integration. For the sake of brevity, we only treat the ball problem. The shell problem can be treated by the same method; see [5].

Let $p_D(t,x,y)$ be the transition density function of the Brownian motion killed at time τ_D. The existence of $p_D(t,x,y)$ is proved in [4]. In fact, we have

$$p_D(t,x,y) = p(t,x,y) - E^x[p(t - \tau_D, X_{\tau_D}, y);\ t < \tau_D],$$

where $p(t,x,y) = (2\pi t)^{-d/2}\exp\{-\|y - x\|^2/2t\}$ is the transition density function of the (free) Brownian motion on R^d. When D is bounded and smooth, $p_D(t,x,y)$ can also be defined as the unique minimal solution of the heat equation with Dirichlet boundary condition:

$$(1)\begin{cases} \frac{\partial}{\partial t}p_D(t,x,y) = \frac{1}{2}\Delta_y p_D(t,x,y), & t > 0,\ x \in D,\ y \in D; \\ p_D(t,x,y) = 0 & t > 0,\ x \in D,\ y \in \partial D; \\ \lim_{t\to 0} p_D(t,x,y) = \delta_x(y), & x \in D,\ y \in D; \end{cases}$$

where δ_x is the Dirac delta function at x.

THEOREM 1. Let D be a bounded domain of C^3 boundary and $x \in D$. We have

$$(2) \qquad P^x[\tau_D \in dt, X_{\tau_D} \in dy] = \tfrac{1}{2}\frac{\partial p_D(t,x,y)}{\partial n_y}dt\sigma(dy),$$

where n_y is the inward normal direction at $y \in \partial D$ and σ is the $(d - 1)$ dimensional volume measure on ∂D.

PROOF. Let f be a nonnegative continuous function on ∂D and $\alpha > 0$. Define

$$(3) \qquad u(x) = E^x[e^{-\alpha\tau_D} f(X_{\tau_D})].$$

Then u is the unique solution of the Dirichlet problem

$$\begin{cases} (\frac{\Delta}{2} - \alpha)u = 0, & \text{on } D; \\ u = f & \text{on } \partial D; \end{cases}$$

see [2]. One the other hand,

$$(4) \qquad G_\alpha(x,y) = \int_0^\infty e^{-\alpha t} p_D(t,x,y)dt$$

is Green's function of the first kind for the operator $\frac{\Delta}{2} - \alpha$ on D. Thus by Green's representation formula, we have

$$(5) \qquad u(x) = \tfrac{1}{2}\int_{\partial D} \frac{\partial G_\alpha(x,y)}{\partial n_y} f(y)\sigma(dy).$$

From (3), (4) and (5) it follows that

$$(6)\quad E^x[e^{-\alpha\tau_D} f(X_{\tau_D})] = \int_0^\infty e^{-\alpha t} \int_{\partial D} \frac{1}{2} \frac{\partial p_D(t,x,y)}{\partial n_y} f(y)\sigma(dy)dt.$$

It can be verified that under our assumptions on the domain, the integral converges absolutely and the exchange of derivation and integration needed in deriving (6) is legitimate. Formula (2) now follows at once from (6) by inverting the Laplace transform.

We now specialize the situation by taking D to be the unit ball B centered at the origin. To apply Theorem 1, we need to compute the transition density function $p_B(t,x,y)$ explicitly. In the following, we will assume that $d \geq 3$. The same method is applicable to the cases d = 1 and 2, but the final formulas look slightly different. We use $J_\nu(r)$ to denote the Bessel function of order ν and use $C_m^\nu(t)$ to denote the Gegenbauer polynomial. The latter is defined via its generating function:

$$(7)\qquad (1 - 2\alpha t + \alpha^2)^{-\nu} = \sum_{n=0}^\infty C_n^\nu(t)\alpha^n.$$

We set $q = (d - 2)/2$.

THEOREM 2. _The transition density function_ $p_B(t,x,y)$ _for the killed Brownian motion in the unit ball_ $B = \{x \in R^d : \|x\| < 1\}$ _is equal to_

$$\frac{2}{q\omega_{d-1}}(\|x\|\|y\|)^{-q}$$

$$\sum_{\substack{m\geq 0\\ n\geq 0}} (q + m)\frac{J_{m+q}(\mu_{n,m}\|x\|)}{J'_{m+q}(\mu_{n,m})} \frac{J_{m+q}(\mu_{n,m}\|y\|)}{J'_{m+q}(\mu_{n,m})} C_m^q(\cos\theta)e^{-\mu_{n,m}^2 t/2},$$

<u>where</u> ω_{d-1} = <u>the</u> (d - 1)-<u>dimensional volume of</u> ∂B, θ <u>is the angle</u> xOy <u>and</u> $\{\mu_{n,m}, n > 0\}$ <u>are nonnegative zeros of</u> J_{m+q} <u>in the ascending order</u>.

PROOF. This is a standard exercise in the separation variable technique of mathematical physics. We only indicate the main steps. Fix $x \in B$ and choose a spherical coordinate system $y = (r, \theta_1, \ldots, \theta_{d-2}, \phi)$ so that $x = (\|x\|, 0, \ldots, 0)$. The volume element is $dy = r^{d-1}\sin^{d-1}\theta_1 \ldots \sin\theta_{d-2}\,dr\,d\theta_1 \ldots d\theta_{d-2}\,d\phi$. We regard x as fixed. By symmetry, $p_B(t,x,y)$ is a function of $(t,r,\theta) = (t, \|y\|, \measuredangle xOy)$. (Note that $\theta = \theta_1$). It follows from (1) that $Q(t,r,\theta) = p_B(t,x,y)$ is the solution of

$$(9)\begin{cases} \dfrac{\partial Q}{\partial t} = \dfrac{1}{2}\dfrac{\partial^2 Q}{\partial r^2} + \dfrac{d-1}{2r}\dfrac{\partial Q}{\partial r} + \dfrac{1}{2r^2\sin^{d-2}\theta}\dfrac{\partial}{\partial \theta}\left(\sin^{d-2}\theta\dfrac{\partial Q}{\partial \theta}\right); \\ Q(t,1,\theta) = 0; \\ \lim_{t \to 0} Q(t,r,\theta) = \delta_{(\|x\|,0)}(r,\theta). \end{cases}$$

This equation can be solved by the standard separation variable technique. Let $Q = T(t)R(r)\Theta(\theta)$, we have

$$\frac{dT}{dt} + \frac{\mu^2}{2}T = 0;$$

$$r^2\frac{d^2R}{dr^2} + (d-1)r\frac{dR}{dr} + [\mu^2 r^2 - m(m+d-2)]R = 0;$$

$$\frac{d}{d\theta}\left(\sin^{d-2}\theta\,\frac{d\Theta}{d\theta}\right) + m(m+d-2)\sin^{d-2}\theta\cdot\Theta = 0.$$

The solutions which are meaningful to our problem are

$$T = e^{-\mu_{n,m}^2 t/2},$$

$$R = r^{-q} J_{m+q}(\mu_{n,m} r);$$

$$\Theta = C_m^q(\cos\theta);$$

see [3], p. 971 p. 1031. The completeness of the system

$$(10) \qquad r^{-q} J_{m+q}(\mu_{n,m} r) C_m^q(\cos\theta), \quad n > 0, \ m > 0$$

follows because we can recover the Poisson kernel from this system, a fact which will be proved at the end of this note. Now we seek a representation of the form

$$(11) \quad Q(t,r,\theta) = \sum_{n>0,m>0} B_{n,m} r^{-q} J_{m+q}(\mu_{n,m} r) C_m^q(\cos\theta) e^{-\mu_{n,m}^2 t/2}.$$

The $B_{n,m}$'s can be determined by multiplying (11) with $r^{-q} J_{m+q}(\mu_{n,m} r) C_m^q(\cos\theta)$ and integrating over B. Using the last condition in (9), and with the help of [3], we find that

$$B_{n,m} = A_{n,m} \|x\|^{-q} J_{m+q}(\mu_{n,m} \|x\|),$$

where

$$(12) \quad A_{n,m} = C_m^q(1) \left(\int_0^1 r J_{m+q}^2(\mu_{n,m} r)\,dr\right)^{-1} \left(\int_{\partial B} C_m^q(\cos\theta)^2 \sigma(dy)\right)^{-1} = \frac{2(q+m)}{q\omega_{d-1} [J'_{m+q}(\mu_{n,m})]^2}.$$

Substituting this in (11), we obtain (8).

Combining Theorem 1 and Theorem 2 we have

THEOREM 3. *The joint density function of* (τ_B, X_{τ_B}) *with respect to* $dt\sigma(dy)$ *is*

$$(13)\quad -\frac{1}{q\omega_{d-1}}\|x\|^{-q}$$

$$\sum_{n>0,m>0}(q+m)\mu_{n,m}e^{-\mu_{n,m}^2 t/2}\frac{J_{m+q}(\mu_{n,m}\|x\|)}{J'_{m+q}(\mu_{n,m})}C_m^q(\cos\theta).$$

The next theorem was first proved in [5].

THEOREM 4. *Let* θ_t *be the angle* xOX_t. *We have*

$$E^x[e^{-\alpha\tau_B}C_m^q(\cos\theta_{\tau_B})] = C_m^q(1)\|x\|^{-q}\frac{I_{m+q}(\sqrt{2\alpha}\|x\|)}{I_{m+q}(\sqrt{2\alpha})},$$

where $I_{m+q}(r)$ *is the Bessel function of imaginary argument.*

PROOF. We have

$$E^x[e^{-\alpha\tau_B}C_m^q(\cos\theta_{\tau_B})]$$

$$= \int_0^\infty\int_{\partial D} e^{-\alpha t}C_m^q(\cos\theta)P^x[\tau_B \quad dt, X_{\tau_B} \quad dy].$$

Using (12) and (13), we get

$$(14)\quad E^x[e^{-\alpha\tau_B}C_m^q(\cos\theta_{\tau_B})]$$

$$= -2C_m^q(1)\|x\|^{-q}\sum_{n>0}\frac{\mu_{n,m}}{\mu_{n,m}^2+2\alpha}\frac{J_{m+q}(\mu_{n,m}\|x\|)}{J'_{m+q}(\mu_{n,m})}.$$

But the last sum is just the partial fractional expansion of the meromorphic function $I_{m+q}(\sqrt{2\alpha})\|x\|)/I_{m+q}(\sqrt{2\alpha})$; namely,

$$(15) \quad \frac{I_{m+q}(\sqrt{2\alpha}\|x\|)}{I_{m+q}(\sqrt{2\alpha})} = -2 \sum_{n>0} \frac{\mu_{n,m}}{\mu_{n,m}^2 + 2\alpha} \frac{J_{m+q}(\mu_{n,m}\|x\|)}{J'_{m+q}(\mu_{n,m})}.$$

The theorem is proved.

Finally, we prove that the Poisson kernel P(x,y) for B can be recovered from the explicit density formula (13). As we pointed out early, this implies that the system (10) is complete on B. Integrating (13) from 0 to ∞, we obtain

$$P(x,y) = P^x[X_{\tau_B} \in dy]/\sigma(dy)$$

$$= -\frac{2}{q\omega_{d-1}}\|x\|^{-q} \sum_{n>1,m>0} \frac{q+m}{\mu_{n,m}} \frac{J_{m+q}(\mu_{m,n}\|x\|)}{J'_{m+q}(\mu_{n,m})} C_m^q(\cos\theta).$$

In (15), letting $\alpha = 0$, we have

$$-2 \sum_{n>1} \frac{1}{\mu_{m,n}} \frac{J_{m+q}(\mu_{m,n}\|x\|)}{J'_{m+q}(\mu_{m,n})} = \|x\|^{m+q}.$$

Put this in the expression from P(x,y). By (7), we have

$$P(x,y) = \frac{1}{q\omega_{d-1}} \sum_{m>0} (q+m)\|x\|^m C_m^q(\cos\theta)$$

$$= \frac{1}{\omega_{d-1}}(1 - 2\|x\|\cos\theta + \|x\|^2)^{-q}$$

$$+ \frac{1}{q\omega_{d-1}} \frac{d}{d\|x\|}(1 - 2\|x\|\cos\theta + \|x\|^2)^{-q}$$

$$= \frac{1}{\omega_{d-1}} \frac{1 - \|x\|^2}{(1 - 2\|x\|\cos\theta + \|x\|^2)^{d/2}}.$$

References

[1] Aizeman, M. and Simon, B., Brownian Motion and Harnack's Inequality for Schrödinger Operators, Comm. Pure and Appl. Math., 35 (1982) p. 209-271.

[2] Chung, K. L., Lectures from Markov Processes to Brownian Motion, Springer-Verlag, New York, 1980.

[3] Gradshteyn, I. S. and Ryuzhik, I. M., Table of Integrals, Series and Products, Academic Press, 1979.

[4] Port, S. C., and Stone, C. J., Brownian Motion and Classical Potential Theory, Academic Press, 1978.

[5] Wendel, J. G., Hitting Spheres with Brownian Motion, Ann. of Prob., vol. 8, No. 1 (1980), p. 164.

Pei Hsu
Courant Institute of Mathematical Sciences
251 Mercer Street
New York, New York 10012

Seminar on Stochastic Processes, 1985
Birkhäuser, Boston, 1986

ON THE DURATION OF THE LONGEST EXCURSION

by

F. B. Knight

1. Statement of problem and results

Let X_t be a standard Markov process, and let 0 be a point regular of 0, so that there exists a continuous additive functional $A(t)$ whose points of increase are contained in $\{t : X_t = 0\}$, called the local time at 0. Then $A(t)$ is unique up to a constant factor which is of no relevance to the present paper (see [1] for a more complete discussion and references). We are concerned here with the durations of the excursions away from 0, which does not depend on the normalization of $A(t)$. For arbitrary $t > 0$, we set $t_0(t) = \sup\{s < t : X_s = 0\}$, $t_1(t) = \inf\{s > t : X_s = 0\}$, and $d(t) = t_1(t) - t_0(t)$. Then we call $d(t)$ the duration of the excursion from 0 containing t (we always assume $X_0 = 0$, hence $d(t)$ is well-defined when we permit $d(t) = \infty$).

In the present work we seek to obtain explicitly the distributions of the two quantities $D(t) = \sup\{d(s) : s < t\}$ and $E(t) = \sup\{d(s) : s < t, t_1(s) < t\}$, in other words the maximal duration of excursions starting by time t, or respectively, ending by time t. Our interest in these quantities arose from the observation that they seem to be less well understood than the corresponding suprema of the height of the excursions (assuming that X_t is real valued). The distribution of the maximal height reduces, in the case of excursions starting by time t, to finding the joint distribution of M(t) and X_t where $M(t) = \max\{X_s : s < t\}$, in a rather evident way. Methods for treating this problem are very well understood, based on the hitting probability distributions of intervals $[y,\infty)$ and the strong Markov property. In the case of excursions completed by time t an analogous method can be used. The problem is, perhaps, to find the joint distribution of the time and value of M(t), which at least is a familiar problem. For the maximal duration, on the other hand, we were unaware of any general method until we obtained an expression for the Laplace transforms. We then discovered that equivalent expressions are known in other contexts, going back to [5, (1965)] in the context of Markov random sets. These are limited to the case of D(t), but our method is to obtain the case of E(t) from that of D(t). Consequently, apart from the fact that our proof seems to be new, the only possible contributions of the present paper are

a) to show how the Laplace transforms may be inverted by reduction to an integral equation,

b) to obtain the case of E(t) from that of D(t), and

c) to carry out the inversion explicitly, insofar as we are able, in the special cases of Brownian motion, the Ornstein-Uhlenbeck velocity processes, and the continuous state branching processes.

Being desirous of "quitting while ahead" we have not attempted any further inversions, but the method seems perfectly straightforward. On the other hand, the distribution of D(t) or E(t) for all t is not obtained in a single step, but requires a recursive procedure. For example, one first obtains the distribution of D(t) in $t \leq x \leq \infty$, then in $t/2 \leq x < t$, and in general the solution in $t/n+1 \leq x < t/n$ depends on those obtained already. In no case have we found an explicit expression valid for all n.

Before going into details, we present the expressions obtained.

DEFINITION 1.1. Let $F(t,x) = 1 - P\{D(t) \leq x\} = P\{$the first excursion of duration $> x$ starts by $t\}$, and let $G(t,x) = 1 - P\{E(t) \leq x\} = P\{$the first excursion of duration $> x$ is complete by $t\}$.

THEOREM 1.1. a) Let B_t be standard Brownian motion starting at 0. Then $F(t,x) = F(tx^{-1})$ and $G(t,x) = G(tx^{-1})$ depend only on tx^{-1}. We have[N.B.]

$$F(y) = \begin{cases} 2\pi^{-1}y^{1/2}; & y \leq 1 \\ \pi^{-1}(3 - y + 4/\sqrt{3}\ \ln y); & 1 \leq y \leq 2. \end{cases}$$

(We have not pursued it beyond $y = 2$, where $F(2) \sim .828$).

As a consequence, we also obtain

$$G'(y) = \begin{cases} (y\pi)^{-1}(y-1)^{1/2}; \ 1 < y < 2, \text{ with } G(1) = 0, \\ (y\pi)^{-1}((y-1)^{-1/2} + 4/3((y-1)^{1/2} - y + 1) \\ \qquad + 2/\sqrt{3}\, y^{-1}\ln(\frac{\sqrt{y}-1}{\sqrt{y}+1})); \end{cases}$$

$2 < y < 3$, with $G(2) = 2\pi^{-1} - 2^{-1}$, from which $G(y)$ is obtained by integration for $y < 3$.

b) Let X_t be an Ornstein-Uhlenbeck process with generator $cd^2/dx^2 - bx\, d/dx$, $b > 0$, $c > 0$, $X_0 = 0$.

Then we have

$$\frac{\partial}{\partial t}F(t,x) = 2b\pi^{-1}e^{bt}((e^{2bt} - 1)(e^{2bx} - 1))^{-1/2}; \ 0 < t < x,$$

whence $F(t,x) = 2\pi^{-1}(e^{2bx} - 1)^{-1/2}(bt + \ln(1 + (1 - e^{-2bt})^{1/2})$, $0 < t < x$, and $G(t,x) = 0$, $G(t,t) = 0 = G_t(t,t)$. For $x < t < 2x$, $\frac{\partial^2 G(t,x)}{\partial x \partial t} = -2b^2\pi^{-1}[(1 - e^{-2bx})^3 (e^{-2bx} - e^{-2bt})]^{-1/2}$, from which G is obtained by integration.

c) Let X_t be a diffusion either with generator $c|x|^{-\beta}d^2/dx^2$; $-1 < \beta$, $c > 0$, $X_0 = 0$, $-\infty < x < \infty$, or with generator

$$c(\frac{d^2}{dx^2} + \beta((2+\beta)x)^{-1}\frac{d}{dx}), \ c > 0, \ 0 < x < \infty \text{ reflected at } 0,$$

or with generator

N. B. The expression $F(y)$, $y \le 1$, follows easily from a formula of P. Lévy [8, Theorem 44.4].

$c(\frac{xd^2}{dx^2} + (1 - (2 + \beta)^{-1})\frac{d}{dx})$, $c > 0$, $0 < x < \infty$ as above at 0.

Then $F = F(tx^{-1})$ and $G = G(tx^{-1})$ as in a), and $F(y) = (\sin(\pi(2 + \beta)^{-1})(2 + \beta)\pi^{-1}y^{1/(2+\beta)}$; $y < 1$, while $G'(y) = (\sin \pi(2 + \beta)^{-1})(\pi y)^{-1}(y - 1)^{1/(2+\beta)}$; $1 < y < 2$, with $G(y) = 0$ for $y < 1$.

NOTES. 1. Cases b) and c) reduce to a) when the parameter b or β is 0. Neither result depends on c (> 0), as will be discussed below.

2. In the "next" interval, such as $1 < y < 2$ in case c), the integration apparently does not simplify. In case c), for instance, our best expression for $F(y)$, setting $\alpha = (2 + \beta)^{-1}$, is

$$F(y) = (\sin \pi\alpha)\pi^{-1}[\alpha^{-1}(y - 1)^{\alpha}$$

$$+ (\sin \pi\alpha)\pi^{-1}\int_0^{y-1} (y - 1 - x)^{\alpha-1} \int_x^1 (1 + x - v)^{-1-\alpha}v^{\alpha}dvdx]$$

and $G(y)$ is doubtless even more complicated $2 < y < 3$. Our policy in the present paper has been to state only the results for which the integration has been carried out in closed form.

3. Cases b) and c) suggest that one could use the observed duration of excursions to estimate the parameters b and β. We have made some statistical investigation in this direction, and the outcome seems worth mentioning, particularly in case c). To remark first on the method, it is to be noted that $F(t,x)$ and $G(t,x)$ are changed only

nominally by two kinds of transformations. First, change of t to kt in the process simply replaces $F(t,x)$ by $F(k^{-1}t, k^{-1}x)$, and multiplies the generator by k^{-1}. This can easily be checked in b) above. Second, and perhaps more important, both F and G are invariant under arbitrary continuous change of the scale x, provided that either 0 is fixed or we replace it by its image. This is clear because the duration of excursions does not involve the space variable directly, and it explains the three alternatives in case c). A consequence of these invariances is already seen in the fact that neither the result of b) nor that of c) depends on the parameter labelled c. Such invariance may be either an advantage or a disadvantage to the estimation, depending on what is being estimated. It would seem to be an advantage for estimating the parameters which are invariant. Another remark would be that, since the observations involve only the arrival times at a fixed point, they can be used as check on, or supplement to, spatially oriented observations. Indeed, the usual statistical tests for a diffusion require calculating the "quadratic variation" along the path in order to obtain the diffusion coefficient before estimating the drift ([LeBreton and Musiela, 6] for example, and [Rubin and Tucker, 10] in the context of Lévy processes). This would seem to require much more elaborate measurement than the present approach. On the other hand at least in the diffusion case the present method does not fall too much short of determining the entire process uniquely. If we can assume that 0 is a reflecting barrier, then it follows from the inverse spectral theorem of M. G. Krein that either the Lévy measure of $\{0\}$, or even just the arrival times at 0 except for a P-null set, determine the entire speed measure $m(dx)$, (including any Feller boundary conditions at the other endpoint) uniquely up to a scale factor. This remarkable fact is discussed in [3]. Consequently, we need to ask to what extent $F(x,t)$ determines the Lévy measure. It is very easy to show, for example, that $F(x,t)$, $0 < t < \infty$, for each x determines the Lévy measure in $[0,x]$ uniquely up to a constant factor. In particular, when $F = F(x/t)$, as in case c) above, $F(x,t)$ for any fixed x or fixed t determines the Lévy measure up to linear change of scale. Of course in cases b) or c) we do not have this much, since F is only computed in an interval. There, however, the problem is simply estimation of a parameter which alone determines the Lévy measure uniquely up to a scale factor.

This estimation, while routine in most respects, presents a certain novelty in the question of what to do with the observations which fall outside the interval in which F is computed. Letting X denote an observation and I_θ the interval in which F_θ is known, (θ = the parameter), we can calculate $p_\theta = P_\theta\{X \in I_\theta\}$, and then the number of observations falling outside I_θ gives an estimate of p_θ, hence of θ. Thus it would be wasteful to ignore these observations while, on the other hand, since those inside I_θ give an estimate of θ while those outside tend to estimate p_θ, the two kinds do not combine very conveniently to estimate either θ or p_θ, (in particular, we cannot solve the combined maximum likelihood equations for either θ or p_θ explicitly). In this situation, our response would be the following. Suppose we continue the observations until a prescribed number k fall inside interval I_θ, which in our case turns out to be free of θ so that the number n of trials is a stopping time. Then, for every θ, n is independent of the k observations in I_θ, and negative binomial. We can use the maximum likelihood estimator k/n of p_θ independently of the estimator based on the observations in I_θ. It is left to the reader to combine the two in any sensible way, for example by using a linear combination which minimizes the maximum variance.

We present these results first in case c). The observations which seem most useful here are those obtained by fixing x and observing the starting time of the first excursion of duration exceeding x. We take for convenience x = 1, and then the distribution of the starting time is $p_\beta y^{1/2+\beta}$, $0 < y < 1$, where $p_\beta = (\sin \pi/(2+\beta))((2+\beta)/\pi)$. It is most convenient to take $\theta = 2 + \beta$, $2 < \theta$, as the parameter. Then the conditional density of observations in (0,1) is $\theta^{-1}t^{-1+1/\theta}$. Instead of basing our estimate on t directly, however, it is in order to use $-\ln t$, $0 < t < 1$. Indeed, if T has the above density then $-\ln T$ has an exponential density with mean θ. Thus our prescription for estimating θ, based on the first k observations $T_1, T_2, \ldots, T_k$ falling in (0,1), is simply to use the sample mean $\bar{\theta} = -1/k \sum_{j=1}^{k} \ln T_j$. This is at once the maximum likelihood estimate, the minimum variance unbiased estimate, and also the unique minimum risk equivariant estimate for loss functions of the form $L(\theta,\hat{\theta}) = |1 - \hat{\theta}/\theta|^\gamma$, $0 < \gamma$, [7, pp. 174-175]), when θ^{-1} is considered as a scale parameter. On the other hand, our independent estimate of p_β by k/n is also well understood. Here, since p_β ranges monotonically from $2/\pi$ to 1 as β increases, p_β is perhaps large enough that one may afford to ignore the observations not falling in (0,1) and use the estimator $\bar{\theta}$ alone.

Returning to $\bar{\theta}$, it is worth making a comparison with the corresponding "spatial" estimator in the case of the "branching with immigration" generator $xd^2/dx^2 + (1 + \beta)/(2 + \beta)\, d/dx$, with 0 a reflecting barrier. Here the process at t = 1, starting at 0, has density $\Gamma^{-1}(b)e^{-y}y^{b-1}$ where $b = (1 + \beta)/(2 + \beta)$ (see for example, [4, 4.3, §5]). Its mean and variance are both b, while those of $-\ln T$ are θ and θ^2 respectively. It seems surprising that the spatial observations have a Γ-density, while the temporal ones have an exponentially distributed logarithm. Our guess is that this similarity should not detract from the fact that entirely different quantities are observed, and that neither observation should be a very good substitute for the other.

Turning to the Ornstein-Uhlenbeck family b), it again seems the best use of Theorem 1.1 (computationally and otherwise) to base the estimation on the starting time of the first excursion of length exceeding a fixed x (the ending time would involve G(t,x) which is more complicated, while by fixing t and using the duration as a random variable one has the problem that the duration may be arbitrarily large). In this case, however, the estimation of parameter b is much more difficult. A better way to base an estimate of the Lévy measure would no doubt be to simply wait for the first excursion of duration exceeding

ε, for some fixed $\varepsilon > 0$. Its duration has distribution $1 - n(x,\infty)/n(\varepsilon,\infty)$, $\varepsilon < x$, so it can provide estimates of $n(x,\infty)/n(\varepsilon,\infty)$ and hence of b (I am indebted to Professor R. M. Blumenthal for this remark).

2. Proofs and derivations.

We first prove a representation theorem for the Laplace transform

$$L_\beta(x) = \int_0^\infty e^{-\beta t} F(t,x)\,dt, \quad \beta > 0.$$

An equivalent form of the result is in [5], but without use of local time it is in a rather primitive state. It is well known (for example [1]) that the right-continuous inverse $A^{-1}(\alpha)$ of the local time at 0 is an increasing process with homogeneous, independent increments, and therefore that

$$E \exp(-\beta A^{-1}(\alpha)) = \exp(-\alpha(m_0\beta + \int_0^{\infty+} (1 - e^{-\beta y})n(dy)))$$

where $0 < m_0$ and $n(y,\infty]$ is the Lévy measure, with $n\{\infty\} = 0$ if 0 is recurrent.

THEOREM 2.1.

$$\beta L_\beta(x) = [e^{-\beta x} + \frac{\beta}{n(x,\infty]}(m_0 + \int_0^x e^{-\beta y} n(y,\infty]dy)]^{-1}$$

<u>if</u> $n(x,\infty] \neq 0$.

PROOF. We begin by recalling that $P\{D(A^{-1}(\alpha)) < x\} = P\{$no excursion starting by time $A^{-1}(\alpha)$ has duration exceeding $x\} = e^{-\alpha n(x,\infty]}$. This is because the excursions exceeding x form a Poisson process in the local time A(t) with parameter $\lambda = n(x,\infty]$ -- a well-known consequence of the Ito excursion theory. Now let us set

$$A_x^{-1}(\alpha) = \int_0^{A^{-1}(\alpha)} I_{\{t_1(t)-t_0(t)<x\}}(t)dt,$$

namely the total duration of excursions starting by time $A^{-1}(\alpha)$ and of length at most x. Then we have

$$\{D(t) < x\} \subset (\{A_x^{-1}(A(t)) > t\} \cap S) \cup N,$$

where $S = \{t$ does not either end or begin an excursion$\}$ and $N \subset S^c$. Now since the exceptional set of t is countable, it is not hard to see that $P(S) = 1$, and so $P(N) = 0$. Indeed, $P\{t$ begins an excursion$\} = 0$ by the Markov property, while

$$P\{t \text{ ends an excursion}\} = E\int_0^t n\{t - s\}dA(s)$$

$$= \int_0^t n\{t - s\}d(EA(s)) = 0$$

because EA(s) is continuous.

For fixed n, set

$$S_{k,n}(t) = \{A_x^{-1}(k2^{-n}) < t < A_x^{-1}((k + 1)2^{-n})\}.$$

Then we have

$$S_{k,n}(t) \cap \{D(t) < x\} \subset (S_{k,n}(t) \cap \{D(A_x^{-1}(k2^{-n})) < x\}),$$

while

$$S_{k,n}(t) \cap \{D(A_x^{-1}((k + 1)2^{-n})) < x\} \subset (S_{k,n}(t) \cap \{D(t) < x\}).$$

Now we recall from general excursion theory that the process $A_x^{-1}(\alpha)$, which depends only on excursions of duration at most x by time $A^{-1}(\alpha)$, is entirely independent of $A^{-1}(\alpha) - A_x^{-1}(\alpha)$, which depends only on excursions exceeding x. Moreover, since $P\{A^{-1}(k2^{-n})$ terminates an excursion exceeding $x\} = 0$, we have a.s.

$$\{D(A_x^{-1}(k2^{-n})) < x\} = \{D(A^{-1}(k2^{-n})) < x\}$$

$$= \{A^{-1}(k2^{-n}) - A_x^{-1}(k2^{-n}) = 0\}.$$

Consequently, the above inclusions imply that

$$\exp(-(k + 1)2^{-n}n(x,\infty])P(S_{k,n}) < P(S_{k,n}(t) \cap \{D(t) < x\})$$

$$< \exp(-k2^{-n}n(x,\infty])P(S_{k,n}(t)).$$

Therefore,

$$\exp(-2^{-n}n(x,\infty]) \sum_k (\exp(-k2^{-n}n(x,\infty])P(S_{k,n}(t)) < P\{D(t) < x\}$$

$$< \sum_k (\exp(-k2^{-n}n(x,\infty])P(S_{k,n}(t))).$$

Then by dominated convergence it follows that

$$\int_0^\infty e^{-\alpha n(x,\infty]} d_\alpha \left(\int_0^\infty e^{-\beta t} P\{t < A_x^{-1}(\alpha)\} dt\right)$$

$$= \lim_{n\to\infty} \sum_k (\exp(-k2^{-n} n(x,\infty])$$

$$\int_0^\infty e^{-\beta t} P\{A_x^{-1}(k2^{-n}) < t < A_x^{-1}(k+1)2^{-n}\} dt)$$

$$= \lim_{n\to\infty} \int_0^\infty e^{-\beta t} \sum_k (\exp(-k2^{-n} n(x,\infty] P(S_{k,n}(t)))dt$$

$$= \int_0^\infty e^{-\beta t} P\{D(t) < x\} dt = \frac{1}{\beta} - L_\beta(x).$$

On the other hand the left side is

$$\int_0^\infty e^{-\alpha n(x,\infty]} d_\alpha \left(\frac{1}{\beta} + \frac{1}{\beta}\int_0^\infty e^{-\beta t} d_t P\{t < A_x^{-1}(\alpha)\}\right)$$

$$= -\beta^{-1} \int_0^\infty e^{-\alpha n(x,\infty]} d_\alpha E \exp(-\beta A_x^{-1}(\alpha))$$

$$= -\beta^{-1} \int_0^\infty e^{-\alpha n(x,\infty]} d_\alpha \exp\left(-\alpha\left(m_0\beta + \int_0^x (1 - e^{-\beta y}) n(dy)\right)\right)$$

$$= \left(\frac{\beta^{-1}(m_0\beta - (1 - e^{-\beta x}) n(x,\infty] + \beta\int_0^x e^{-\beta y} n(y,\infty] dy)}{n(x,\infty] + m_0\beta - (1 - e^{-\beta x}) n(x,\infty] + \beta\int_0^x e^{-\beta y} n(y,\infty] dy}\right)$$

$$= \beta^{-1}\left(1 - n(x,\infty]\left(m_0\beta + e^{-\beta x} n(x,\infty] + \beta\int_0^x e^{-\beta y} n(y,\infty] dy\right)^{-1}\right).$$

Consequently, we have

$$\beta L_\beta(x) = n(x,\infty]\left(e^{-\beta x} n(x,\infty] + m_0\beta + \beta\int_0^\infty e^{-\beta y} n(y,\infty] dy\right)^{-1}$$

which reduces easily to the required expression.

COROLLARY 2.2. *For each* x, $L_\beta(x)$ *determines* $F(t,x)$, $t > 0$, *uniquely, and this uniquely determines* $(m_0;\ n(y,\infty],\ 0 < y < x)$ *up to a constant factor*.

PROOF. We saw before that P{t starts an excursion} = 0, so $F(t,x)$ is continuous in t. Hence it is uniquely determined by inversion of $L_\beta(x)$. Now we write Theorem 2.1 in the form

$$((\beta L_\beta(x))^{-1} - e^{-\beta x})\beta^{-1} = (n(x,\infty])^{-1}(m_0 + \int_0^x e^{-\beta y} n(y,\infty]dy).$$

Then as $\beta \to \infty$ the right side has limit $(n(x,\infty])^{-1}m_0$. Subtracting this from both sides determines $(n(x,\infty])^{-1}\int_0^x e^{-\beta y} n(y,\infty]dy$, from which $n(y,\infty](n(x,\infty])^{-1}$ is determined by inversion in β.

COROLLARY 2.3. *We have*

$$G(t,x) = \int_0^{t-x} n(x,t - s]/n(x,\infty]F(ds,x) \text{ for } t > x,$$

and $G(t,x) = 0$ *for* $t < x$. *In particular, for each* x *we determine* $G(t,x)$ *from* $F(s,x)$ *and* $n(y,\infty]/n(x,\infty]$, $x < y < t$.

PROOF. We first show that for any t and $0 < \varepsilon < x < t$, we have the double inequality

$$\frac{n(x,t]}{n(x,\infty]}P\{D(\varepsilon) > x\} \le P\{D(\varepsilon) > x \text{ and } t_1(\varepsilon) \le t\}$$

$$\le \frac{n(x,t + \varepsilon]}{n(x,\infty]}P\{D(\varepsilon) > x\}.$$

Indeed, since $D(\varepsilon) > x$ if and only if $d(\varepsilon) > x$, we have

$$P\{x < d(\varepsilon) \le t\} \le P\{D(\varepsilon) > x \text{ and } t_1(\varepsilon) \le t\}$$

$$\le P\{x < d(\varepsilon) \le t + \varepsilon\}.$$

Now setting $\alpha_x = A(\inf\{s : d(s) > x\})$, it follows by Ito excursion theory that for $c > x$

$$P\{d(A^{-1}(\alpha_x) -) \le c \,|\, \mathscr{A}_{\alpha_{\bar{x}}}\} = \frac{n(x,c]}{n(x,\infty]}$$

over $\{\inf\{s : d(s) > x\} < \infty\}$, where $\mathscr{A}_t$ denotes the natural filtration of the process (subordinator) $A^{-1}(\alpha)$. This is an immediate application of M. Weil's formula for conditioning given the strict past [12, Theorem 1], since $d(A^{-1}(\alpha_x)-)$ exceeds x on the above set, where

$$\alpha_x = \inf\{\alpha : A^{-1}(\alpha) - A^{-1}(\alpha-) > x\} < \infty.$$

Therefore, we have $\{d(\varepsilon) > x\} = \{A^{-1}(\alpha_{\bar{x}}) < \varepsilon\} \in \mathscr{A}_{\alpha_{\bar{x}}}$, so

$$P\{x < d(\varepsilon) \le t\} = E(P\{d(A^{-1}(\alpha_x)-) \le t \,|\, \mathscr{A}_{\alpha_{\bar{x}}}\};\{A^{-1}(\alpha_{\bar{x}}) < \varepsilon\})$$

$$= \frac{n(x,t]}{n(x,\infty]} P\{d(\varepsilon) > x\}.$$

The proof for the right inequality is analogous.

Next, for $0 \le k < 2^n$, $0 < n$, set $I_{k,n} = (k2^{-n}(t - x), (k + 1)2^{-n}(t - x)]$, introduce the stopping times $T_{k,n} = \inf\{s \in I_{k,n} : X_s = 0\}$, $0 \le k < 2^n$, with $\inf\{\phi\} = \infty$, and write $S_{k,n} = \{\inf\{s : d(s) > x \text{ and } t_1(s) \le t\} \in I_{k,n}\}$.

Then we have

$$G(t,x) = \sum_k P(S_{k,n})$$

$$= \sum_k [P\{(\exists s \in (T_{k,n},(k+1)2^{-n}(t-x)] : d(s) > x \text{ and } t_1(s) < t)$$

$$\cap(\nexists s \in I_{j,n} : d(s) > x \text{ and } t_1(s) < t,\ j < k)].$$

By the strong Markov property, this becomes

$$\sum_k E[P(\exists s \in (T_{k,n},(k+1)2^{-n}(t-x)] : d(s) > x \text{ and}$$

$$t_1(s) < t \mid T_{k,n});\ \nexists s \in I_{j,n} : d(s) > x \text{ and}$$

$$t_1(s) < t,\ j < k]$$

$$= \sum_k E[P(\exists s \in (0,(k+1)2^{-n}(t-x) - T_{k,n}] : d(s) > x \text{ and}$$

$$t_1(s) < t - T_{k,n} \mid T_{k,n});\ \nexists s \in I_{j,n} :$$

$$d(s) > x \text{ and } t_1(s) < t,\ j < k].$$

Now by the start of the proof the conditional probabilities in the last expression are between $n(x,t - T_{k,n}]/(x,\infty]$ $P\{D((k+1)2^{-n}(t-x) - T_{k,n}) > x\}$ and the same expression with $n(x,t - T_{k,n} + \varepsilon]$ (of course, these are 0 if $T_{k,n} = \infty$). Then the corresponding sums can be written

$$\sum_k E\Big(\frac{n(x,t - T_{k,n}]}{n(x,\infty]} P(\exists\, s \in (T_{k,n}, (k+1)2^{-n}(t - x)] :$$

$$d(s) > x \,|\, T_{k,n}); \;\exists\, s \in I_{j,n} : d(s) > x \text{ and}$$

$$t_1(s) < t,\; j < k\Big),$$

and the same expression with $n(x,t - T_{k,n} + \varepsilon]$. On the other hand, over the set $\{\exists\, s \in I_{j,n} : t_1(s) > t,\; j < k\}$ we have $T_{k,n} = \infty$, so we can just as well take the expectations over $\{\exists\, s \in I_{j,n} : d(s) > x,\; j < k\}$. Since

$$k2^{-n}(t - x) < T_{k,n} < (k + 1)2^{-n}(t - x),$$

it follows that our original sum, for each n, is bounded below and above by

$$\sum_k \Big(\frac{n(x,t - (k + 1)2^{-n}(t - x)]}{n(x,\infty]}$$

$$E[P(\exists\, s \in (T_{k,n}, (k + 1)2^{-n}(t - x)] : d(s) > x \,|\, T_{k,n}; \exists$$

$$s \in I_{j,n} : d(s) > x,\; j < k]\Big)$$

and the corresponding sum with $n(x,t - k2^{-n}(t - x)]$. But here the expectation is simply $F((k + 1)2^{-n}(t - x)) - F(k2^{-n}(t - x))$, so the lower sum may be written as

$$\int_0^{t-x} \frac{n(x,t - s_n^+)]}{n(x,\infty]} F(ds,x),$$

and the upper sum as

$$\int_0^{t-x} \frac{n(x,t - s_n^-(s)]}{n(x,\infty]} F(ds,x),$$

with $s_n^-(s) < s < s_n^+(s)$, $s_n^+(s) - s_n^-(s) = 2^{-n}(t - x)$. Moreover, $s_n^{\pm}(s)$ are monotone in n, hence by monotone convergence we have

$$\int_0^{t-x} \frac{n(x,t - s)}{n(x,\infty]} F(ds,x) < G(t,x) < \int_0^{t-x} \frac{n(x,t - s]}{n(x,\infty]} F(ds,x).$$

Finally, since $F(t,x)$ is continuous in t, this reduces to the assertion of Corollary 2.3.

We next show that inversion of $L_\beta(x)$ in Theorem 2.1 reduces to solving a Volterra integral equation of convolution type.

THEOREM 2.4. *For each* $x > 0$, $t > 0$, $F(s,x)$ *for* $0 < s < t$ *is the unique bounded continuous solution of the equation*

$$(2.1) \quad t = \int_{0-}^{t} \Big(I_{(x,\infty)}(s) + \frac{m_0\delta_0(s)}{n(x,\infty]} + \frac{n(s,\infty]}{n(x,\infty]} I_{(0,x]}(s)\Big)F(t - s,x)ds,$$

where $\delta_0(s)$ *is the unit mass at* 0.

REMARK. This is an equation of the second type if $m_0 \neq 0$, of the first type if $m_0 = 0$.

PROOF. Since the kernel is not bounded near 0, standard existence and uniqueness theorems do not apply. The method is simply to show the equivalence of this equation to the identity of Theorem 2.1. We begin by writing this last in the form

$$\beta^{-1} = \beta L_\beta(x)[\frac{e^{-\beta x}}{\beta} + \frac{1}{n(x,\infty]}(m_0 + \int_0^x e^{-\beta y}(n(y,\infty]dy)].$$

The left side is the Laplace transform of dt, while the term in brackets on the right is the transform of the measure

$$\mu(dt) = I_{(x,\infty)}(t)dt + \frac{m_0}{n(x,\infty]}\delta_0(t) + (\frac{n(t,\infty]}{n(x,\infty]}I_{(0,x)}(t))dt.$$

Since $L_\beta(x)$ is the transform of $F(t,x)dt$, which is continuous with $F(0,x) = 0$, the right side may be written as

$$\beta\int_0^\infty e^{-\beta t}(\int_0^t F(t - s,x)\mu(ds))dt = \int_0^\infty e^{-\beta t}d(\int_0^t F(t - s,x)\mu(ds)).$$

Therefore, $dt = d(\int_0^t F(t - s,x)\mu(ds))$, and the equation of Theorem 2.4 follows by integration. Conversely, the identity of these measures with a bounded continuous function F implies the equation in the Laplace transforms, and this determines F uniquely by Corollary 2.2. Hence the proof is complete.

We have solved (2.1) only in cases where $m_0 = 0$. Then it is expedient first to solve the equation for $t < x$, where it becomes

$$(2.2) \qquad t\, n(x,\infty] = \int_0^t n(s,\infty]F(t - s,x)ds.$$

For the cases in point, this will be a generalized Abel equation whose explicit solution is not difficult. Once we know $F(s,x)$, $s < x$, we can continue to $x < t < 2x$ by writing (2.1) in the form

$$t\, n(x,\infty] - \int_{t-x}^{x} n(t - s,\infty]F(s,x)ds - n(x,\infty] \int_{0}^{t-x} F(s,x)ds$$
$$= \int_{x}^{t} n(t - s,\infty]F(s,x)ds,$$

with the left side a known function. Setting $F_1(s,x) = F(s + x,x)$, and $u = t - x$, the right side becomes

$$\int_{0}^{u} n(u - s,\infty]F_1(s,x)ds; \qquad 0 < u < x,$$

which gives an equation of the same form (2.2) but a new left side. In the same way, if $F(s,x)$ is known for $s < nx$, we can reduce it to this form for $nx < t < (n + 1)x$ by setting $F_n(s,x) = F(s + nx,x)$, $u = t - nx$, to obtain the equation for $0 < u < x$

$$G_n(u) = \int_{0}^{u} n(u - s,\infty]F_n(s,x)ds, \tag{2.3}$$

where

$$G_n(u) = (u + nx)n(x,\infty] - \int_{(n-1)x+u}^{nx} n(u + nx - s,\infty]F(s,x)ds$$
$$- n(x,\infty] \int_{0}^{(n-1)x+u} F(s,x)ds.$$

REMARK. In case $m_0 \neq 0$ in (2.1), the same methods can be applied. For $s < x$ we have

$$(2.5)\qquad t\, n(x,\infty] = m_0 F(t,x) + \int_0^t n(s,\infty]F(t-s,x)ds,$$

and by induction on n

$$(2.6)\quad G_n(u) = m_0 F_n(u,x) + \int_0^u n(u-s,\infty]F_n(s,x)ds,\ 0 < u < x,$$

where $F_n(s,x) = F(s + nx,x)$, and $G_n(u)$ is given by (2.4).

We now turn to deriving the solutions for $F(t,x)$ in the cases of Theorem 1.1. We combine cases a) and c), since a) reduces to c) with $\beta = 0$ (except that we are able to continue a) for an additional interval). Since our calculations do not depend on the exact normalization of the Lévy measure $n(x,\infty]$, we will write $g(x) = cf(x)$ when $f(x)$ is known and c is a constant which varies from place to place. For the diffusion with generator $c|x|^{-\beta}d^2/dx^2$ on (∞,∞), $-1 < \beta$, it is known [2, Section 6.7] that

$$\int_0^\infty (1 - e^{-\lambda s})n(ds) = c\lambda^\alpha$$

where $\alpha = (2 + \beta)^{-1}$ and from this it is easy to check that $n(ds) = cs^{-(\alpha+1)}ds$, and $n(x,\infty] = cx^{-\alpha}$. Thus our equation (2.2) reduces to

$$(2.7)\qquad tx^{-\alpha} = \int_0^t (t-s)^{-\alpha}F(s,x)ds.$$

This is a generalized Abel equation [11, Section 1.12], whose solution is obtained by compostion with the kernel $(y - t)^{\alpha-1}$. Thus we obtain

$$(2.8)\quad F(t,x) = x^{-\alpha}\,\frac{\sin\alpha\pi}{\pi}\,\frac{d}{dt}\int_0^t \frac{s}{(t-s)^{1-\alpha}}\,ds$$

$$= \frac{\sin\alpha\pi}{\pi}\,x^{-\alpha}\,\frac{d}{dt}\int_0^t \frac{t-u}{u^{1-\alpha}}\,du$$

$$= \frac{\sin\alpha\pi}{\pi}\,x^{-\alpha}\int_0^t u^{\alpha-1}\,du = \frac{\sin\alpha\pi}{\alpha\pi}\left(\frac{t}{x}\right)^{\alpha},$$

as asserted in Theorem 1.1 c). We note that since $n(s,\infty](n(x,\infty])^{-1}$ depends only on sx^{-1}, we can write $F(t-s,x) = F((t-s)x^{-1})$ in (2.1) and a change of variables $y = sx^{-1}$ reduces (2.1) to the case $x = 1$ with tx^{-1} in place of t. Thus, in general, the solution depends only on tx^{-1}.

For the interval $1 < t < 2$ (with $x = 1$) the equation to be solved (setting $c_\alpha = \sin\alpha\pi/\alpha\pi$) is

$$t = c_\alpha\left(\int_0^{t-1} v^\alpha dv + \int_{t-1}^{1} (t-v)^{-\alpha} v^\alpha dv\right) + \int_1^t (t-v)^{-\alpha} F(v)\,dv,$$

or letting $F_1(v) = F(v+1)$ as before, we get

$$(2.9)\quad (u+1) - c_\alpha\left(\int_0^u s^\alpha ds + \int_u^1 (u+1-s)^{-\alpha} s^\alpha ds\right)$$

$$= \int_0^u (u-s)^{-\alpha} F_1(s)\,ds.$$

Now composition with the kernel $(y-u)^{\alpha-1}$, and noting that $c_\alpha\int_0^1 (1-s)^{-\alpha} s^\alpha ds = 1$, yields

$$F_1(y) = \frac{\sin\alpha\pi}{\pi}\int_0^y \Big(1 - c_\alpha\,\frac{d}{du}\Big(\int_0^u s^\alpha ds$$

$$+ \int_u^1 (u+1-s)^{-\alpha} s^\alpha ds\Big)(y-u)^{\alpha-1} du$$

$$= \frac{\sin \pi\alpha}{\pi}\int_0^y \left(1 + \frac{\sin \pi\alpha}{\pi}\int_u^1 (u + 1 - s)^{-(\alpha+1)} s^\alpha ds\right)(y - u)^{\alpha-1} du,$$

(as in [11, (8), Section 1.12]) and this verifies Note 2 of Theorem 1.1 (replacing y by y - 1 to get F).

Unfortunately this integral seems to be untractible, even in the Brownian case $\alpha = \frac{1}{2}$. In that case, however, a fresh start leads to a much simpler expression. We begin by setting $u + 1 - s = v$ in the second integral of (2.9):

$$(2.10)\quad (u + 1) - \frac{2}{\pi}\left(\frac{2}{3}u^{3/2} + \int_u^1 (u + 1 - v)^{1/2} v^{-1/2} dv\right)$$

$$= \int_0^u s^{-1/2} F_1(u - s) ds.$$

Now composition with $(y - u)^{-1/2}$ gives

$$(2.11)\quad \int_0^y (y - u)^{-1/2} (u + 1) du - \frac{2}{\pi}\int_0^y (y - u)^{-1/2} \left(\frac{2}{3}u^{3/2}\right.$$

$$\left. + \int_u^1 (u + 1 - v)^{1/2} v^{-1/2} dv\right) du$$

$$= \pi \int_x^y F_1(u) du.$$

Interchanging y - u and u on the left, and differentiating, gives by routine integrations,

$$(2.12)\quad \pi F_1(y) = \frac{d}{dy}\left[\int_0^y (y - u + 1) u^{-1/2} du - \frac{2}{\pi}\int_0^y \left(\frac{2}{3}(y - u)^{3/2}\right.\right.$$

$$\left.\left. + \int_{y-u}^1 (y - u + 1 - v)^{1/2} v^{-1/2} dv\right) du\right]$$

$$= y^{-1/2} + \int_0^y u^{-1/2} du - \frac{2}{\pi}\int_0^y (y - u)^{1/2} u^{-1/2} du$$

$$- \frac{2}{\pi} \int_0^1 (1 - v)^{1/2} v^{-1/2} dv \; y^{-1/2}$$

$$+ \frac{2}{\pi} \int_0^y (y - u)^{-1/2} u^{-1/2} du$$

$$- \frac{1}{\pi} \int_0^y \left(\int_{y-u}^1 (y - u + 1 - v)^{-1/2} v^{-1/2} dv \right) u^{-1/2} du$$

$$= y^{-1/2} + 2y^{1/2} - y - y^{-1/2} + 2$$

$$- \frac{2}{\pi} \int_0^y \sin^{-1}\left(\frac{1 + u - y}{1 + y - u}\right) u^{-1/2} du$$

$$= 2 + 2y^{1/2} - y$$

$$- \frac{2}{\pi} \int_0^y \sin^{-1}\left(\frac{1 - v}{1 + v}\right) (y - v)^{-1/2} dv.$$

This might appear to be the final word, but it is not. Differentiating in y the next to last line yields, by using [9, Formula 154] after some obvious simplifications,

$$(2.13) \quad \pi F_1'(y) = -1 + \frac{4}{\pi} \int_0^y \left(1 - \left(\frac{1 + u - y}{1 + y - u}\right)^2\right)^{-1/2}$$

$$(1 + y - u)^{-2} u^{-1/2} du$$

$$= -1 + \frac{4}{\pi} \int_0^y ((1 + y - u)^2$$

$$- (1 + u - y)^2)^{-1/2} (1 + y - u)^{-1} u^{-1/2} du$$

$$= -1 + \frac{2}{\pi} \int_0^y \frac{1}{\sqrt{yu - u^2}} (1 + y - u)^{-1} du$$

$$= -1 + \frac{4}{\pi(y + 1)} \int_{-1}^1 \frac{1}{\sqrt{1 - v^2}} \left(\frac{1}{2 - v}\right) dv$$

$$= -1 + \frac{4}{\pi\sqrt{3}(y + 1)}$$

Now since $F_1(0) = 2/\pi$ we obtain

$$F_1(y) = \frac{2}{\pi} + \frac{1}{\pi}\int_0^y\left(-1 + \frac{4}{\sqrt{3}(y + 1)}\right)dy = \frac{2}{\pi} - \frac{y}{\pi} + \frac{4}{\sqrt{3}\pi}\ln(y + 1).$$

It seems remarkable that such a simplification in (2.12) occurs, but extending it to $\alpha > 1/2$ remains an open problem.

Next we consider the Ornstein-Uhlenbeck family of Theorem 1.1 b). The first problem is to obtain the Lévy measures $n(x,\infty)$ up to a constant factor (this is presumably known but we do not know of a reference). According to a well-known formula [2, 6.2, 2)] the Lévy measure is connected with the Green function $G_\beta(0,0)$ by

$$(2.14) \qquad c = G_\beta(0,0)\left(m_0 + \int_0^\infty(1 - e^{-\beta y})n(dy)\right)$$

(this does not depend, of course, on any particular scale for the process). The transition density at 0, up to a constant factor, is $\sqrt{b}(1 - e^{-2bt})^{-1/2}$, so we obtain (since $m_0 = 0$ and $\int_0^\infty(1 - e^{-\beta y})n(dy) = \beta\int_0^\infty e^{-\beta y}n(y,\infty)dy$)

$$(2.15) \quad c = \left(\beta\int_0^\infty e^{-\beta y}n(y,\infty),dy\right)\left(\sqrt{b}\int_0^\infty e^{-\beta t}(1 - e^{-2bt})^{-1/2}\,dt\right).$$

On the other hand, setting $e^{-t} = v$ we have

$$(2.16) \quad \int_0^\infty e^{-\beta t}(1 - e^{-t})^{-1/2}\,dt = \int_0^1 v^{\beta-1}(1 - v)^{-1/2}\,dv$$

$$= \frac{\sqrt{\pi}\Gamma(\beta)}{\Gamma(\beta + 1/2)}$$

so it follows that by (2.15) that

$$(2.17)\quad \beta \int_0^\infty e^{-\beta y} n(y,\infty)dy = cb^{1/2}\,\Gamma^{-1}(\tfrac{\beta}{2b})\Gamma(\tfrac{\beta}{2b} + \tfrac{1}{2})$$

$$= cb^{1/2}\,\Gamma^{-1}(\tfrac{\beta}{2b})\Gamma(\tfrac{\beta}{2b} - \tfrac{1}{2})(\tfrac{\beta}{2b} - \tfrac{1}{2}),$$

or again

$$(2.18)\quad \int_0^\infty e^{-\beta y} n(y,\infty)dy = c(b^{-1/2} - \frac{b^{1/2}}{\beta})(\Gamma^{-1}(\tfrac{\beta}{2b})\Gamma(\tfrac{\beta}{2b} - \tfrac{1}{2})).$$

But again by (2.16), the last factor is the transform of $1/\sqrt{\pi}(1 - e^{-t})^{-1/2}$ at $(\beta/2b - 1/2)$, i.e. the transform of $2b/\sqrt{\pi}e^{bt}(1 - e^{-2bt})^{-1/2}$ at β, hence by inversion we obtain

$$(2.19)\quad n(t,\infty) = c\sqrt{b}\,e^{bt}(1 - e^{-2bt})^{-1/2}$$

$$- b^{3/2}\int_0^t e^{bs}(1 - e^{-2bs})^{-1/2}\,ds$$

$$= c[\sqrt{b}\,e^{bt}(1 - e^{-2bt})^{-1/2}$$

$$- \sqrt{b}\,e^{bt}(1 - e^{-2bt})^{1/2}]$$

$$= c\sqrt{b}\,e^{-bt}(1 - e^{-2bt})^{-1/2}.$$

Now for $t < x$, equation (2.1) gives

$$(2.20)\quad t = (1 - e^{-2bx})^{1/2}e^{bx}$$

$$\int_0^t e^{-bs}(1 - e^{-2bs})^{-1/2} F(t - s,x)ds.$$

To solve this, we again take Laplace transforms, ignoring the restriction to $t < x$. This gives (now denoting by $L(\beta,x)$ the transform of extended $F(t,x)$)

$$(2.21)\quad \beta^{-2} = (1 - e^{-2bx})^{1/2} e^{bx}$$

$$\int_0^\infty e^{-(b+\beta)s}(1 - e^{-2bs})^{-1/2} ds\ L(\beta,x)$$

$$= (e^{2bx} - 1)^{1/2} \frac{\sqrt{\pi}}{2b}\Gamma(\tfrac{\beta}{2b} + \tfrac{1}{2})\Gamma^{-1}(\tfrac{\beta}{2b} + 1)L(\beta,x),$$

from which it follows that

$$(2.22)\quad L(\beta,x) = \frac{1}{\sqrt{\pi}}(e^{2bx} - 1)^{-1/2} \beta^{-1}(\Gamma(\tfrac{\beta}{2b})\Gamma^{-1}(\tfrac{\beta}{2b} + \tfrac{1}{2})).$$

The last factor is the transform of $2b\pi^{-1/2}(1 - e^{-2bt})^{-1/2}$ in view of (2.16) again, and inverting $L(\beta,x)$ to obtain $F(t,x)$ in $t < x$ yields

$$(2.23)\quad F(t,x) = \frac{2b}{\pi}(e^{2bx} - 1)^{-1/2} \int_0^t (1 - e^{-2bs})^{-1/2} ds.$$

Differentiating in t gives the first expression of Theorem 1.1 b), and a short table of integrals then gives the second one (after change of variables).

REMARK. This result can also be obtained from the corresponding case of Theorem 1.1 a), by representing the process as a non-random time change of Brownian motion.

The method breaks down, however, for the next interval $x < t < 2x$, where it seems that $n(t,\infty)$ from (2.19) must be used, but the calculations are formidable.

It remains only to calculate $G(t,x)$ in the three cases, using Corollary 2.3. For example, in case c) we obtain easily

$$G(y) = \pi^{-1}\sin\left(\frac{\pi}{2+\beta}\right)\int_0^y (1-(y-z)^{-\frac{1}{2}+\beta})z^{-(1+\beta)/(2+\beta)}dz,$$

and the assertion follows by a routine differentiation for the case $t < x$. The proof for case b) is entirely analogous.

Finally, in the interval $2 < y < 3$ for G, in the Brownian case, we have

$$G(y) = \frac{1}{\pi}\Big[\int_0^1 (1-(y-u)^{-\frac{1}{2}})u^{-\frac{1}{2}}\,du$$

$$+\int_1^{y-1}(1-(y-u)^{-\frac{1}{2}})\left(\frac{4}{\sqrt{3}}u^{-1}-1\right)du\Big],$$

and routine differentiation together with tedious integration (including $\int x^{-1}(1-x)^{-\frac{1}{2}}dx$ from an integral table) leads finally to the asserted expression.

FINAL REMARK. Another method for $x < t < 2x$, also applicable in the general case, is to combine G and F by observing that $F(t,x) = G(t,x) + P\{d(t) > x\} - P\{$the first excursion of duration $> x$ is complete by t and $d(t) > x\}$. By Lévy's result for $B(t)$ cited in Theorem 1.1 a), Brownian excursion $(t_0(t),t_1(t))$ has joint density

$$(2\pi)^{-1}(t_0(t_1 - t_0)^3)^{-1/2}; \quad 0 < t_0 < t < t_1.$$

From this it follows by writing

$$X_t \equiv e^{-bt}B(\tfrac{1}{2b}(e^{2bt} - 1))$$

(see, for example, [4, 4.3, §3]) that the corresponding density for the Ornstein-Uhlenbeck process X_t is

$$2b^2\pi^{-1}e^{2b(t_0+t_1)}((e^{2bt_0} - 1)(e^{2bt_1} - e^{2bt_0})^3)^{-1/2}$$

(I am indebted to Professor K. Burdzy for this suggestion). Hence we can compute $d/dx\, P\{d(t) > x\}$, and it emerges that

$$G(t,x) + P\{d(t) > x\}$$

$$= \frac{2}{\pi}(e^{2bx} - 1)^{-1/2}(bt + \ln(1 + (1 - e^{-2bt})^{1/2})),$$

$$x < t < 2x.$$

In other words, the formula for $F(t,x)$, $t < x$, is the same as that for $G(t,x) + P\{d(t) > x\}$ when $x < t < 2x$ (this is also true for $B(t)$, and seems to be a general fact). The third term, however, is complicated. It can be written

$$\int_x^t \partial/\partial s\, G(s,x)F(t - s,x)ds$$

where the integrand is known from Theorem 1.1 b), but the integration apparently does not simplify.

References

1. B. Fristedt and S. J. Taylor, Constructions of local time for a Markov process, Z. Wahrsheinlichkeitstheorie verw. Gebiete 62 (1983), 73-112.

2. K. Ito and H. P. McKean, Jr., Diffusion Processes and their sample paths, Academic Press, New York, 1965.

3. F. B. Knight, Characterization of the Lévy measures of inverse local time of gap diffusion, Progress in Prob. Statist. 1, Birkhauser, Boston, Mass., 1981, 53-78.

4. F. B. Knight, Essentials of Brownian motion and diffusion, Math. Surveys, 18, Amer. Math. Soc., Providence, R. I., 1981.

5. N. Krylov and A. A. Juskevich, Markov random sets, Trans. Moscow Math. Soc. 13 (1965), 127-152.

6. A. Le Breton and M. Musiela, A study of one-dimensional model for stochastic processes, Prob. Math. Statist. 2, Fasc 1 (1984), 91-107.

7. E. L. Lehmann, Theory of Point Estimation, J. Wiley and Sons, New York, 1983.

8. P. Lévy, Processus Stochastiques et Mouvement Brownien, Gauthier-Villars, 1948.

9. B. O. Peirce, A Short Table of Integrals, Ginn and Co., Boston, 1914.

10. H. Rubin and H. Tucker, Estimating the parameters of a differential process, Ann. Math. Statist. 30 (1959), 641-658.

11. F. G. Tricomi, Integral Equations, Interscience Pub. Inc., New York, 1957.

12. M. Weil, Conditionnement par rapport au passé strict, Seminaire de Prob. V, Springer Lecture Notes 191, Springer Verlag, (1971), 362-372.

F. B. Knight
Department of Mathematics
University of Illinois
Urbana, Illinois 61801

Seminar on Stochastic Processes, 1985
Birkhäuser, Boston, 1986

STRICT PAST CONDITIONING AT ARBITRARY TIMES

by

B. Maisonneuve

1. Introduction

Let $X = (\Omega, \mathcal{F}, \mathcal{F}_t, X_t, \theta_t, P^x)$ be a Hunt process with state space $(E, \mathcal{E})$. In [4] Weil proved the conditional independence of $\mathcal{F}_{T-}$ and θ_T given X_{T-} for certain hitting times T of the process $(X_{t-}, X_t)_{t>0}$. In [3] Pitman investigated the stochastic dependence of $\mathcal{F}_{T-}$ and θ_T for arbitrary stopping times T. Here we shall look at this question for arbitrary times T. The main result is formula (3) below which contains both Weil's result, its extension to arbitrary hitting times of (X_{t-}, X_t) (Theorem 2) and a result for last exit times. This was thought to be an introduction to similar formulae for the excursions straddling arbitrary times, but we had no time to work out the details for the present publication.

2. Notations

Let $D = \{t > 0 : X_{t-} \neq X_t\}$. In [1] Benveniste and

Jacod established the existence of a Lévy system (n,A) for X. With the notation $N^x = \int_E n(x,dy)P^y$, $x \in E$ one has the following formula

$$(1)\qquad P^\mu \sum_{s\in D} Z_s 1_B(k_s,s,\theta_s) = P^\mu \int_{\mathbb{R}_+} dA_s Z_s N^{X_s}(B_{k_s,s})$$

for each initial distribution μ, for all positive predictable processes Z and all $B \in \mathscr{F}^o \otimes \mathscr{B}_{\mathbb{R}_+} \otimes \mathscr{F}$. As usual k_t, $t > 0$ denote the killing operators and $\mathscr{F}^o = \sigma(X_t,\ t > 0)$. In the right hand side of (1) $N^{X_s}(B_{k_s,s})$ denotes the mapping $\omega \to N^{X_s(\omega)}(B_{k_s\omega,s})$, with the notation

$$B_{\omega,s} = \{\omega' \in \Omega : (\omega,s,\omega') \in B\}.$$

Let T be a fixed random time on $(\Omega,\mathscr{F})$. For all $\omega \in \Omega$ consider the set

$$(2)\qquad C^\omega = \{T(\omega|T(\omega)|\cdot) = T(\omega)\},$$

with the familiar notation $\omega|t|\omega'$ for the map w which agrees with ω on $[0,t[$ and satisfies $\theta_t\omega = \omega'$ if $t < \infty$.

3. The basic formula

THEOREM 1. <u>For</u> $0 < T(\omega) < \infty$, <u>we set</u> $\nu^\omega = N^{X_{T-}(\omega)}$. <u>Then for</u> P^μ a.e. ω <u>such that</u> $T(\omega) \in D(\omega)$.

(i) <u>the set</u> C^ω <u>belongs to the</u> ν^ω <u>completion of</u> $\mathscr{F}^o$ <u>and</u> $0 < \nu^\omega(C^\omega) < \infty$,

(ii) for all $B \in \mathscr{F}^{\circ}$

(3) $$P^{\mu}(\theta_T \in B | \mathscr{F}_{\overline{T}-})(\omega) = \nu^{\omega}(B|C^{\omega})$$

with the notation $\nu(B|C) = \nu(BC)/\nu(C)$ if $0 < \nu(C) < \infty$, 0 otherwise and with $\overline{T} = T$ on $\{T \in D\}$, $+\infty$ on $\{T \notin D\}$.

PROOF. Let T', T" be $\mathscr{F}^{\circ}$ measurable times such that $T' \leq T \leq T''$ and $P(T'' \neq T') = 0$ and set $C = \{(\omega,t,\omega') : T'(\omega|t|\omega') \leq t \leq T''(\omega|t|\omega')\}$. For every positive predictable Z, with $Z_{\infty} = 0$, one has

$$Z_{\overline{T}} 1_B(\theta_T) = \sum_{s \in D} Z_s 1_B(\theta_s) 1_C(k_s,s,\theta_s) \text{ a.s.}$$

and it follows from (1) that

$$P^{\mu}(Z_{\overline{T}} 1_B(\theta_T)) = P^{\mu}\int_{\mathbb{R}_+} dA_s Z_s N^{X_{s-}}(BC_{k_s,s}).$$

By the argument of Theorem 1 of Weil [4] (see Pitman [3] for a systematic use of this argument), we obtain

a) $0 < N^{X_{T-}}(BC_{k_T,T}) < \infty$ a.s. on $\{\overline{T} < \infty\} = \{T \in D\}$.

b) $P^{\mu}(Z_{\overline{T}} 1_B(\theta_T)) = P^{\mu}(Z_{\overline{T}} N^{X_{T-}}(B|C_{k_T,T}))$.

Since $P^{\mu}(T'' \neq T') = 0$, it follows from b) and a monotone class argument that

$$N^{X_{T-}}(\{T''(k_T|T|\cdot) \neq T'(k_T|T|\cdot)\}C_{k_T,T}) = 0$$

a.s. on $\{T \in D\}$. Thus the mapping $\omega \to \nu^{\omega}(C^{\omega})$ is P^{μ} a.s. defined and is $\mathscr{F}^{\mu}$ measurable; i), ii) now follow from a), b).

4. Applications

1) Suppose that $T = \inf\{t > 0 : (X_{t-}, X_t) \in H\}$, where H is a nearly Borel set for the processes $(X_{t-}, X_t)_{t>0}$ and $(X_{t-}, X_{t-})_{t>0}$. We define

$$F = \{x : P^x(T = 0) = 1\}, \; K = \{x : (x,x) \in H\} \cup F.$$

The following result is an extension of Theorem 1 of Weil [4].

THEOREM 2. <u>With the previous notations one has</u>

$$(5) \quad P^{\mu}(\theta_T \in B | \mathscr{F}_{T-}) = Q^{X_{T-}}(B) \text{ on } \{0 < T < \infty\},$$

where $Q^x(B) = P^x(B)$ if $x \in K$, $N^x(B | (x, X_0) \in H \text{ or } T = 0)$ <u>if</u> $x \notin K$.

PROOF. By proposition 3 of Weil [4], the set $\{T \in D\}$ is in $\mathscr{F}_{T-}$ and by applying (3) we get $P^{\mu}(\theta_T \in B | \mathscr{F}_{T-})(\omega)$ $= \nu^{\omega}(B | C^{\omega})$ for a.e. $\omega \in \{T \in D\}$. In the present situation one has $C^{\omega} = \{(X_{T-}(\omega), X_0) \in H \text{ or } T = 0\}$ whenever $0 < T(\omega) < \infty$. By using the following lemma, we obtain (5).

LEMMA. $\{T \in D\} = \{0 < T < \infty, X_{T-} \notin K\}$ <u>a.s.</u>

PROOF. It follows from (1) that

$P^{\mu}(X_{T-} \in K, T \in D)$

$$= P^{\mu} \sum_{s \in D} 1_{\{s<T, X_{s-} \in K\}} 1_{\{X_{s-}, X_s) \in H \text{ or } T \circ \theta_s = 0\}}$$

$$= P^{\mu} \int_{[0,T]} dA_s 1_{\{(X_{s-}, X_{s-}) \in H \text{ or } X_{s-} \in F\}} U_s,$$

where $U_s(\omega) = N^{X_{s-}(\omega)}((X_{s-}(\omega), X_0) \in H \text{ or } T = 0)$. But (A_t) being continuous, this last term equals

$$P^{\mu} \int_{[0,T[} dA_s 1_{\{X_{s-}, X_s) \in H \text{ or } X_s \in F\}} U_s = 0. \; *$$

REMARKS. This lemma extends a known result for Hunt processes (see Dellacherie [2], p. 72: if A is nearly Borel for X and if $T = \inf\{t > 0 : X_t \in A\}$, then $P^{\mu}(X_{T-} \in A, T \in D) = 0$). If H is admissible in the sense of Weil [4] ($(X_{T-}, X_T) \in H$ a.s. on $\{0 < T < \infty\}$), the statement is equivalent to

(6) $\{T \in D\} = \{0 < T < \infty, (X_{T-}, X_{T-}) \notin H\}$ a.s.

and one checks that (5) reduces to Theorem 1 of Weil [4]. Note that the equality (6) was implicitly used by Weil in the proof of his Theorem 1. The argument used for our lemma is a modification of the argument of Proposition 5 of [8], which has been reminded to me by P. Fitzsimmons. I thank him very much.

2) Let H be like in 1) and set

$$S = \sup\{t > 0 : (X_{t-}, X_t) \in H\} \quad (\sup \emptyset = 0),$$

$$T = S \text{ if } 0 < S < \infty \text{ and } (X_{S-}, X_S) \in H, +\infty \text{ otherwise.}$$

In this case $C^{\omega} = \{(X_{T-}(\omega), X_0) \in H,\ S = 0\}$ whenever $T(\omega) < \infty$. Hence, provided there exists K such that $\{T \in D\} = \{0 < T < \infty,\ X_{T-} \notin K\}$ a.s., formula (5) still holds, with a modified Q^x (replace "or T = 0" by " and S = 0"). For instance if H does not intersect the diagonal this condition holds with $K = \emptyset$.

3) Let $A \in \mathcal{P} \otimes \mathcal{F}$, where $\mathcal{P}$ denotes the predictable σ-field on $\Omega \times \mathbb{R}_+$. We assume that for a.e. ω the set $\Delta(\omega) = \{t \in D(\omega) : (\omega, t, \theta_t\omega) \in A\}$ contains at most one point. Then for the random time $T = \inf \Delta (\inf \emptyset = +\infty)$, one has $C^{\omega} = A_{\omega, T(\omega)}$ for a.e. $\omega \in \{T < \infty\}$ (note that for each P^{μ} there exists a set A' such that for P^{μ} a.e. ω one has $\forall t > 0,\ \forall \omega';\ (\omega, t, \omega') \in A \iff (k_t\omega, t, \omega') \in A'$). It is easy to see that formula (3) applied to this case is just another formulation of theorem (2.6) of [3] stated for the $(\Omega, \mathcal{F})$ valued point process $(\theta_t)_{t \in D}$.

* To finish the proof of the lemma, it should also be voted that on $\{0 < T < \infty,\ T \notin D\}$ one has $((X_{T-}, X_{T-}) \in H$ or $T \circ \theta_T = 0)$ or equivalently $X_{T-} \in K$ due to the strong Markov property at T.

References

[1] A. Benveniste and J. Jacod. Systèmes de Lévy des processus de Markov. Invent. Math. 21 (1973), 183-198.

[2] C. Dellacherie. Au sujet des sauts d'un processus de Hunt. Séminaire de Probabilités IV, 71-72, Lecture Notes Math. 124. Springer, Berlin, 1970.

[3] J. W. Pitman. Lévy Systems and path decompositions. Seminar on Stochastic Processes, 1981, 79-110. Birkhäuser, Boston, 1981.

[4] M. Weil. Conditionnement par rapport au passé strict. Séminaire de Probabilités V, 362-372, Lecture Notes Math. 191. Springer, Berlin, 1971.

B. Maisonneuve.
I.M.S.S.
Université de Grenoble II
47X-38040 Grenoble Cedex.
France

Seminar on Stochastic Processes, 1985
Birkhäuser, Boston, 1986

DISCONTINUOUS TIME CHANGES AND DUALITY FOR MARKOV PROCESSES

by

Joanna B. Mitro

1. Introduction

The purpose of this paper is to continue work on discontinuous time changes of dual Markov processes begun in [6], where time changes based on discontinuous **natural** additive functionals were discussed. Both continuous and discontinuous natural time changes share the features

(i) dual time changes (i.e., based on dual additive functionals) preserve duality,

(ii) the new duality measure of the time changed processes is constructed from the Revuz measure of the time changing additive functionals.

In this paper, we consider the general case of time changes based on additive functionals which have a quasi-left-continuous purely discontinuous component, and present for them a time changing procedure which possesses features (i) and (ii) above.

The time change of [6] and the one constructed here are based on Weidenfeld's approach to discontinuous time changes [8]. A different approach is latent in Cinlar and Kaspi's construction of a local time process in [3] (see also [5]); the idea is to replace the jumps of an additive functional with independent, exponential variables, which are independent of the process as well (on an enlarged sample space). The time changed process is created using the inverse of the new functional. We discuss duality for this time change in the last section.

Our basic process $X = (\Omega, \mathscr{F}, \mathscr{F}_t, X_t, \theta_t, (P^x)_{x\in E})$ is a right continuous, strong Markov process on the state space $(E, \mathscr{E})$, which we assume is Lusin. As usual, Δ denotes the cemetery point, ζ the lifetime, (P_t) the transition semigroup, and $(U^\alpha, \alpha > 0)$ the resolvent of the process X. We assume further that X is standard in the sense of [4], and in duality with a standard process $\hat{X}$ on E, relative to the σ-finite measure ξ.

Basic facts concerning the structure of discontinuous additive functionals of dual processes may be found in [7]. We shall use the **bivariate Revuz measure**

(11) $$\nu_A(F) = \lim_{t\to 0} \frac{1}{t} E^\xi \int_{(0,t]} F(X_{s-}, X_s)\, dA_s$$

introduced there. Set $\nu_A^1(\Gamma) = \nu_A(\Gamma \times E)$. For a natural additive functional (NAF) A, ν_A is carried by the diagonal D in $E \times E$; if the NAF A is purely discontinuous, $\nu_A^1 = f d\lambda$, where f is a Borel function which vanishes off some semipolar set and λ is a canonical measure on the σ-ring of semipolar sets. For A purely discontinuous and quasi-left-continuous (p.d.q.l.c.), $\nu_A = Fd\nu$, where F is a Borel function on $E \times E$ which vanishes on D, and ν is a canonical "jumping measure" on $E \times E$. When discussing a generic additive functional A, $\tau_t = \inf\{s : A_s > t\}$ will denote its right continuous inverse and $N = \{u : A_s = u \text{ for some } u\}$ will denote its range.

2. A General Discontinuous Time Change

In this section, our aim is to construct a Weidenfeld type time change, which preserves duality, for additive functionals with a p.d.q.l.c. part. Recall that Weidenfeld's procedure consists of enlarging the sample space and state space in order to accommodate the addition of a decreasing semilinear component, $R_t = A(\tau_t) - \min(t, A_\infty)$, to the usual time changed process $X(\tau_t)$. This time change does not preserve duality, but adding a second, **increasing** semilinear component $S_t = \min(t,\infty) - \sup\{s < t : s \in N\}$ is enough to preserve duality for time changes based on **natural** additive functionals [6]. More is required when the functional has a p.d.q.l.c. part. Consider, for example, a p.d.q.l.c. functional with Revuz measure $Fd\nu$, where F vanishes off $B \times C$, B and C disjoint Borel sets. The state space of the time changed process is a subset of $\mathbb{R}^+ \times \mathbb{R}^+ \times C$, while the time change of the dual process takes values in the **disjoint** region $\mathbb{R}^+ \times \mathbb{R}^+ \times B$. In this example, the additive functional is determined by jumps of the process between the two sets B and C, and the new time change proposed below would produce a process on $\mathbb{R}^+ \times \mathbb{R}^+ \times B \times C$. The extra E-valued component corresponds to the switch to a bivariate Revuz measure.

Fix a finite, perfect additive functional A_t of X (satisfying $A_{t+s} = A_t + A_s \circ \theta_t$ identically). Set $\overline{\Omega} = \mathbb{R}^+ \times \mathbb{R}^+ \times E \times \Omega$. Define the time change of X on $\overline{\Omega}$ by

$$Y_t(\omega) = Y_t(r,s,x,\omega) = (r - t, s + t, x, X_0(\omega)) \qquad \text{if } r > t$$

$$= (R_{t-r}(\omega), S_{t-r}(\omega), X_{(\tau_{t-r})-}(x,\omega), X_{\tau_{t-r}}(\omega)) \text{ if } r < t,$$

where $X(\tau_u-)(x,\omega)$ equals the value of the process $(s,\omega) \to X_{s-}(\omega)$ at $s = \tau_u(\omega)$ provided $\tau_u(\omega) > 0$, and equals x if $\tau_u(\omega) = 0$. On $\overline{\Omega}$ the shift operators $\overline{\theta}_t : \overline{\Omega} \to \overline{\Omega}$ are given by

$$\overline{\theta}_t(r,s,x,\omega) = (r - t, s + t, x, \omega) \qquad \text{if } r > t$$

$$= (R_{t-r}(\omega), S_{t-r}(\omega), X_{(\tau_{t-r})-}(x,\omega), \theta_{\tau_{t-r}}(\omega)) \text{ if } r < t;$$

with this definition, both $\overline{\theta}_u \circ \overline{\theta}_v = \overline{\theta}_{u+v}$ and $Y_t \circ \overline{\theta}_u = Y_{t+u}$ are true (see [6] or [8]). The measures governing Y are $\overline{P}^{(r,s,x,y)} = \varepsilon_r \times \varepsilon_s \times \varepsilon_x \times P^y$ (where ε denotes unit mass). Finally, the filtration $(\overline{\mathscr{F}}_t)$ on $\overline{\Omega}$ comes about by completing in the usual way the field

$$\overline{\mathscr{F}}_t^0 = \{U(r,s,x,\omega) : U \in \mathscr{B}(\mathbb{R}^+) \times \mathscr{B}(\mathbb{R}^+) \times \mathscr{E} \times \mathscr{F},$$

$$U(r,s,x,\cdot)1_{\{t>r\}} \in \mathscr{F}_{\tau_{t-r}}, \ U(r,s,x,\cdot)1_{\{t<r\}} \in \mathscr{F}_0\};$$

$$\overline{\mathscr{F}} = \vee_t \overline{\mathscr{F}}_t.$$

(2.1) PROPOSITION. The process $Y = (\overline{\Omega}, \overline{\mathscr{F}}, \overline{\mathscr{F}}_t, \overline{\theta}_t, Y_t, \overline{P}^{(r,s,x,y)})$ is strong Markov.

PROOF. The proof is similar to the proof of the strong Markov property for Weidenfield's process. We sketch the proof to indicate necessary changes, but see [8] for more detail. We show

$$\int_{\overline{A}} f(Y_{\overline{T}+u})\, d\overline{P}^{(r,s,x,y)} = \int_{\overline{A}} \overline{E}^{Y_{\overline{T}}}(f(Y_u))\, d\overline{P}^{(r,s,x,y)}$$

for $\overline{T}$ an $(\overline{\mathscr{F}}_t^0)$-stopping time, $\overline{A} \in \overline{\mathscr{F}}_{\overline{T}}^0$, and $f \in \mathscr{B}(\mathbb{R}^+) \times \mathscr{B}(\mathbb{R}^+) \times \mathscr{E} \times \mathscr{E}$. Since r, s, and x are fixed, set $A = \{\omega : (r,s,x,\omega) \in \overline{A}\}$, $B = A \cap \{\omega : \overline{T}(r,s,x,\omega) > r\}$, and $T(\omega) = (\overline{T}(r,s,x,\omega) - r) \vee 0$. Then A is the union of the three sets $A_1 = A \cap \{\omega : \overline{T}(r,s,x,\omega) < r\}$, $A_2 = B \cap \{R_T < u\}$, and $A_3 = B \cap \{R_T > u\}$, and it suffices to prove

$$(2.2) \quad \int_{A_j} f(Y_{\overline{T}+u}(r,s,x,\cdot))\, dP^y = \int_{A_j} E^{Y_{\overline{T}}(r,s,x,\cdot)}(f(Y_u))\, dP^y$$

for j = 1 to 3. On A_1, Weidenfield's argument applies. For A_2 and A_3, note that T is an $(\mathscr{F}(\tau_t))$-stopping time, τ_T is an $(\mathscr{F}_t)$-stopping time and $A_2, A_3 \in \mathscr{F}(\tau_T)$. On A_2, $Y_{\overline{T}}(r,s,x,\omega) =$

$$(R_{u-R_T(\omega)}, S_{u-R_T(\omega)}, X_{(\tau_{u-R_T(\omega)})-}(X_{(\tau_T)-}(x,\omega),\cdot),$$

$$X_{\tau_{u-R_T(\omega)}})\theta_{\tau_T}(\omega),$$

and we proceed as Weidenfield does, using the strong Markov property of X at τ_T. On A_3, $R_{T+u} = R_T - u$, $S_{T+u} = S_T + u$, and $\tau_{T+u} = \tau_T$, so the left hand side of (2.2) equals

$$\int_{A_3} f(R_T(\omega) - u, S_T(\omega) + u, X_{(\tau_T)-}(x,\omega), X_0 \circ \theta_{\tau_T}(\omega))P^Y(d\omega)$$

$$= \int_{A_3} P^{X(\tau_T(\omega))}(d\omega')f(R_T(\omega) - u, S_T(\omega) + u, X_{(\tau_T)-}(x,\omega),$$

$$X_0(\omega'))P^Y(d\omega),$$

which is the right hand side for A_3.

In the sequel, we will refer to the process Y as the **full time change** of X to distinguish it from previous time changed processes having fewer components.

3. Duality and the Full Time Change

In this section we prove that the full time change defined in section 2 preserves duality, and compute the new duality measure. The steps of the proof closely parallel the proof of the natural case in [6], and we will not repeat all of the details.

Fix a finite perfect additive functional A of X, and write $A_t = A_t^c + A_t^d$, where A^c is continuous and A^d is purely discontinuous. A_t^d can be written as the sum $A_t^n + A_t^q$, its natural and q.l.c. parts, respectively. In the proof below we begin by showing that the full time change Ψ of X via $B_t = t + A_t^d$ is dual to the full time change $\hat{\Psi}$ of $\hat{X}$ via $\hat{B}$. We finish by obtaining the full time changes Y and $\hat{Y}$ (of X via A and $\hat{X}$ via $\hat{A}$) by time changing Ψ and $\hat{\Psi}$ via dual **continuous** additive functionals. The Revuz measures of A^c,

A^n, and A^q are denoted $\mu(dy)$, $f(y)\lambda(dy)$, and $F(x,y)\nu(dx,dy)$, respectively. The inverse of A is τ_t. In the following statement, note that the components of $\hat{Y}$ (defined in terms of $\hat{A}$) appear in a different order than they do in Y.

(3.1) THEOREM. <u>The processes</u> $Y_t = (R_t, S_t, X(\tau_t-), X(\tau_t))$ <u>and</u> $\hat{Y}_t = (\hat{S}_t, \hat{R}_t, \hat{X}(\tau_t), \hat{X}(\tau_t-))$ <u>are in weak duality relative to the measure</u> $\eta(dr,ds,dx,dy)$ <u>on</u> $\mathbb{R}^+ \times \mathbb{R}^+ \times E \times E$ <u>given by</u>

$$\int h d\eta = \int h(0,0,y,y)\mu(dy) + \int_E \int_{r=0}^{f(y)} h(r, f(y) - r, y, y) dr \lambda(dy)$$

$$+ \int_{E\times E} \int_{r=0}^{F(x,s)} h(r, F(x,y) - r, x, y) dr \nu(dx,dy).$$

PROOF. **STEP 1: Duality of Ψ and $\hat{\Psi}$.** The duality measure γ for Ψ and $\hat{\Psi}$ is obtained by replacing $\mu(dx)$ in the definition of η by $\xi(dx)$. Let W^α, $\hat{W}^\alpha$ denote the resolvents of Ψ and $\hat{\Psi}$, respectively. For $\alpha > 0$ and $h \in b(\mathcal{B}(\mathbb{R}^+) \times \mathcal{B}(\mathbb{R}^+) \times \mathcal{E} \times \mathcal{E})^+$,

$$(3.2) \quad W^\alpha h(r,s,x,y) = 1_{\{r>0\}} \int_0^r e^{-\alpha t} h(r - t, s + t, x, y) dt$$

$$+ e^{-\alpha}(V^\alpha h^{00}(y) + u_D^\alpha(y)),$$

where $(V^\beta, \beta > 0)$ is the resolvent of the subprocess (X,M) for $M_t = \exp(-\alpha A_t^d)$, $h^{00}(z) = h(0,0,z,z)$, and u_D^α is the α-potential of the purely discontinuous additive functional D of (X,M) whose natural and q.l.c. parts are

$$D_t^n = \int_{(0,t]} H_1(X_u) f^{-1}(X_u) M_u dA_u^n$$

$$\left(\tfrac{0}{0} = 0\right)$$

$$D_t^q = \int_{(0,t]} H_2(X_{u-},X_u)M_u dA_u^q,$$

where $H_1(y) = \int_0^{f(y)} \exp(\alpha(f(y) - u))h(f(y) - u,u,y,y)du$ and

$$H_2(x,y) = \int_0^{F(x,y)} \exp(\alpha(F(x,y) - u))h(F(x,y) - u,u,x,y)du.$$

To prove

$$(3.3) \qquad \int W^\alpha h \cdot g \, d\gamma = \int h \cdot \hat{W}^\alpha g \, d\gamma$$

for $h, g \in b(\mathscr{B}(\mathbb{R}^+) \times \mathscr{B}(\mathbb{R}^+) \times \mathscr{E} \times \mathscr{E})^+$, one computes the left hand side of (3.3) using (3.2), and shows this equals the "dual" right hand side. This can be done exactly as in [6]: the left hand side of (3.3) is the sum of five terms (i)-(v) analogous to (I)-(V) of [6], section 6. (The first three terms are expanded to include the part of γ due to the p.d.q.l.c. part of A). The equalities (i) = $(\hat{\text{i}})$, (iv) = $(\hat{\text{iv}})$ follow exactly as in [6]. Using lemma (2.9) of [6], we obtain

$$\nu_D^1(dx) = H_1(x)e^{-\alpha f(x)}\lambda(dx) + \int_E H_2(x,y)e^{-\alpha F(x,y)}\nu(dx,dy).$$

Then since $u_D^\alpha(y) = \int v^\alpha(y,x)\nu_D^1(dx)$ [7], the equalities (iii) = $(\hat{\text{iii}})$, (ii) = $(\hat{\text{v}})$, (v) = $(\hat{\text{ii}})$ are obtained by the same reasoning used for the analogous terms in [6].

STEP 2: Duality of Y and $\hat{Y}$. It suffices to check that the continuous additive functionals needed to time change Ψ to Y and $\hat{\Psi}$ to $\hat{Y}$ are dual, or more precisely, to show that both additive functionals have Revuz measure η. This is done exactly as in [6].

4. The Cinlar-Kaspi Discontinuous Time Change

We now consider a different type of time change, inspired by recent work of Cinlar and Kaspi ([3], [5]). They did not **start** with a discontinuous additive functional, but rather **constructed** one (on and enlarged space), to serve as a "local time" for a fixed homogeneous random subset of $\mathbb{R}^+$. However, their technique easily extends to a method of time changing for general discontinuous additive functions of processes with duals (we use the known structure of the discontinuous parts of such functionals).

We begin by providing a value for X_{0-}. Define $(\overline{\Omega}, \overline{\mathscr{F}}, \overline{P}^{xy}) = (\Omega, \mathscr{F}, P^y) \times (E, \mathscr{E}, \varepsilon_x)$ and extend all objects defined on Ω to $\overline{\Omega}$, with the addition that $X_{0-}(\omega, x) = x$ and the new shift is $\overline{\theta}_t(\omega, x) = (\theta_t \omega, X_{t-}(\omega, x))$.

Now fix a perfect (no exceptional set), finite additive functional A_t with $\Delta A_t = F(X_{t-}, X_t)(F(x,x) = 0$ except for x in some semipolar set). For $t > 0$ and $m > 1$, define (on $\overline{\Omega}$)

$$D(t) = \inf\{s > t : F(X_{s-}, X_s) > 0\}$$

$$S_{m,1} = \inf\{t : \frac{1}{m} < F(X_{t-}, X_t) < \frac{1}{m-1}\}$$

$$S_{m,n+1} = S_{m,n} + S_{m,1} \circ \overline{\theta}_{D(S_{m,n})}.$$

$S_{m,1}$ is a terminal time and $\{S_{m,n}\}_n$ are its iterates; clearly, $\{S_{m,n}\}_{m,n}$ exhausts the jumps of A_t. Set $J_t(m) = \sup\{n : S_{m,n} < t\}$ ($\sup \phi = 0$). It is easy (and left to the reader) to prove:

$$(4.1) \qquad S_{m,n} \circ \bar{\theta}_t + t = S_{m,n+J_t(m)}.$$

The time changed process will be defined on a further augmented sample space. Set $(\tilde{\Omega}, \tilde{\mathscr{F}}, \tilde{P}^{xy}) = (\bar{\Omega}, \bar{\mathscr{F}}, \bar{P}^{xy}) \times (W, \mathscr{H}, P)$ for $(W, \mathscr{H}, P) = (\mathbb{R}^+, \mathscr{B}(\mathbb{R}^+), \lambda)^{\mathbb{N}^+ \times \mathbb{N}^+}$, where $\mathbb{N}^+ = \{1,2,\ldots\}$ and λ is the exponential distribution on $(\mathbb{R}^+, \mathscr{B}(\mathbb{R}^+))$ with mean 1. The coordinate variables on W are denoted $W_{m,n}$, and all objects defined on either $\bar{\Omega}$ or W are extended to $\tilde{\Omega}$ in the obvious way. Next, define

$$\mathscr{H}_t^{\bar{\omega}} = \sigma\{W_{m,n} : n < J_t(m,\bar{\omega}),\ m > 1\} \text{ and}$$

$$\tilde{\mathscr{F}}_t^0 = \{A \in \bar{\mathscr{F}}_t \times \mathscr{H} : 1_A(\bar{\omega},\cdot) \in \mathscr{H}_t^{\bar{\omega}} \quad \text{for every } \bar{\omega} \in \bar{\Omega}\},$$

with $\tilde{\mathscr{F}}_t$ its usual completion (with respect to the $\tilde{P}^{xy}$). On $\tilde{\Omega}$ the shift operators are given by

$$\tilde{\theta}_t(\bar{\omega}, w) = (\bar{\theta}_t\bar{\omega}, \phi_{J_t(\bar{\omega})} w),$$

where $\phi_{J_t} : W \to W$ is given by $W_{m,n} \circ \phi_{J_t} = W_{m,J_t(m)+n}$.

4.2 PROPOSITION. $\tilde{X} = (\tilde{\Omega}, \tilde{\mathscr{F}}, \tilde{\mathscr{F}}_t, \tilde{\theta}_t, X_t, \tilde{P}^{xy})$ <u>is strong Markov</u>.

REMARK: $(\tilde{X},\tilde{\mathscr{F}}_t)$ is almost a strong Markov extension of $(\overline{X},\overline{\mathscr{F}}_t)$ in the sense of [2]; the deficiency is that $(\mathscr{F}_t)$ is not right continuous.

PROOF. This result is proved like Proposition 4.28 in [2]. Only minor changes are necessary because our J_t is left continuous and our $(\tilde{\mathscr{F}}_t)$ is not right continuous. In place of Lemma (4.20) of [2], we have: for $\overline{\omega} \in \overline{\Omega}$ fixed, and T: $W \to [0,\infty]$ a stopping time of $(\mathscr{H}_t^{\overline{\omega}})$, $U \in b\mathscr{H}_t^{\overline{\omega}}$, $Z \in b\mathscr{H}$,

$$E(U\cdot Z \circ \phi_{J_T(\overline{\omega})}) = E(U)E(Z);$$

its proof is an easy modification of the argument in [2]. The proposition follows from this as in [2].

Now, define a new process $\tilde{A}$ on $\tilde{\Omega}$ by setting $\tilde{A}_t = A_t^c + \tilde{A}_t^d$, where A^c is the continuous part of A and

$$\tilde{A}_t^d = \sum_m \sum_n 1_{\{S_{m,n}<t\}} F(X_{S_{m,n}-}, X_{S_{m,n}}) \cdot W_{m,n}.$$

Note that $\tilde{A}_t$ is left continuous, adapted to $(\tilde{\mathscr{F}}_t)$, and increasing. To see that $\tilde{A}_t$ is perfectly additive (for $\tilde{\theta}_t$), note that in computing

$$\tilde{A}_{t+u}^d - \tilde{A}_u^d = \sum_m \sum_n 1_{\{u<S_{m,n}<t+u\}} F(X_{S_{m,n}-}, X_{S_{m,n}}) \cdot W_{m,n},$$

the sum over n can be replaced by a sum starting at $n = J_u(m) + 1$. Changing variables $(k = n - J_u(m))$ and using (4.1), this becomes

$$\sum_m \sum_n 1_{\{0<S_{m,k}\circ\bar{\theta}_u<t\}} F(X_{S_{m,n}-}, X_{S_{m,n}}) \circ \bar{\theta}_u \cdot W_{m,k} \circ \phi_{J_u}$$

$$= \tilde{A}^d_t \circ \tilde{\theta}_u.$$

Moreover, $\tilde{A}$ is finite because A is (this is an easy exercise in a.s. convergence).

To define the time changed process using $\tilde{A}$, set $\tilde{\tau}_t(\tilde{\omega}) = \inf\{s : \tilde{A}_s(\tilde{\omega}) > t\}$, $\tilde{\zeta} = \tilde{A}_\infty$, and $Z_t = X(\tilde{\tau}_t)$. Define $\mathcal{M}^0_t = \tilde{\mathcal{F}}_{\tilde{\tau}_{t-}} \vee \sigma(Z_t)$. Put $\mathcal{M} = \tilde{\mathcal{F}}$, and let $\mathcal{M}_t$ be the usual completion of $\mathcal{M}^0_t$ in $\mathcal{M}$. Shifts σ_t for this process are defined as in [3]:

$$\sigma_t(\bar{\omega}, w) = \begin{cases} \tilde{\theta}_{\tilde{\tau}_t(\bar{\omega},w)}(\bar{\omega}, w) & \text{if } \tilde{\tau}_t \neq S_{m,n} \text{ for any } m,n \\ (\theta_{\tilde{\tau}_t(\bar{\omega},w)}\bar{\omega}, \Psi^{\bar{\omega}}_t w) & \text{if } \tilde{\tau}_t = S_{m,n} \text{ for some } m,n, \end{cases}$$

where $\Psi^{\bar{\omega}}_t$ is such that, on $\{\tilde{\tau}_t = S_{j,k}\}$:

$$W_{m,n} \circ \Psi^{\bar{\omega}}_t = W_{m,n} \circ \tilde{\theta}_{\tilde{\tau}_t} \qquad \text{for } m \neq j$$

$$W_{j,1} \circ \Psi^{\bar{\omega}}_t = W_{j,k} - (F(X_{S_{j,k}-}, X_{S_{j,k}}))^{-1}(t - \tilde{A}_{\tilde{\tau}_t})$$

$$W_{j,n} \circ \Psi^{\bar{\omega}}_t = W_{j,n-1} \circ \tilde{\theta}_{\tilde{\tau}_t}, \qquad n > 2.$$

4.3 THEOREM. <u>The process</u> $Z = (\tilde{\Omega}, \mathcal{M}, \mathcal{M}_t, \sigma_t, Z_t, \tilde{P}^{xy})$ <u>is a right continuous strong Markov process</u>.

PROOF. A proof in case $\Delta A_t = \lambda(X_t)$ for $\lambda \in \mathcal{E}^*$ is given in [3], and there is no problem adapting that proof to our situation.

5. Duality and the Cinlar-Kaspi Time Change

We now return to the question of duality. If a pair of dual processes undergoes a Cinlar-Kaspi time change relative to a pair of dual functionals, will the resulting processes still be duals? We shall work here with discontinuous additive functionals which are natural, to avoid the complication of adding an extra component to the state space. This also enables us to use the original Ω, rather than $\overline{\Omega}$. (To deal with the general case, one modifies the natural solution as we did for Weidenfield's time change in section 3.) For this time change, the situation is very nice: the duality measure is precisely the Revuz measure.

Fix a finite, perfect NAF A of X, $A_t = A_t^c + A_t^d$, and suppose A_t is represented by f (Revuz measure $fd\lambda$). The dual functional is $\hat{A}_t = \hat{A}_t^c + \hat{A}_t^d$, and the time changed processes are denoted Z and $\hat{Z}$. First, we establish that Z can be obtained by a **continuous** time change applied to Σ, the Cinlar-Kaspi time change of X via $B_t = t + A_t^d$. Then, following a now-familiar pattern, we show Σ and $\hat{\Sigma}$ are dual, as are the continuous time changes which yield Z and $\hat{Z}$.

Let β_t denote the inverse of $\tilde{B}_t$, the functional constructed on $\tilde{\Omega} = \Omega \times W$ as in the last section. Define a process C on $\tilde{\Omega}$ by

$$C_t = A_{\beta_t}^c + \int_0^t 1_\Gamma(\Sigma_s)ds,$$

where $\Gamma = \{x : f(x) > 0\}$. It is clear that $C_0 = 0$, $t \to C_t$ is non-decreasing and continuous and $C_t \in \mathcal{M}_t$. The additivity follows from the strong additivity of A and the additivity (relative to σ_t) of β (see [3]).) Thus C is a continuous additive functional. Let ρ denote the inverse of C, and set $Z_s^* = \Sigma_{\rho_s}$.

(5.1) PROPOSITION. <u>The processes</u> Z^* <u>and</u> Z <u>are identical</u>.

PROOF. By definition, $Z_s^* = X(\beta(\rho_s))$ and $Z_s = X(\tilde{\tau}_s)$. We claim $\beta(\rho_s) = \tilde{\tau}_s$. Note that β_t is constant during those intervals when $\Sigma_t \in \Gamma$, and the lengths of such intervals are governed by the jumps of $\tilde{X}$. Outside those intervals, β_t increases linearly (with slope 1). Therefore $A^c \circ \beta$ grows like A^c outside those intervals. What this shows is that when C first surpasses the level s, $\beta \circ \rho$ is recording the time $\tilde{X}$ first exceeds level s. This proves the claim.

The main result of this section is

(5.2) THEOREM. Z <u>and</u> $\hat{Z}$ <u>are in weak duality relative to the Revuz measure of</u> A, $\eta = \mu + f \cdot \lambda$, <u>where</u> μ <u>is the Revuz measure of</u> A^c.

PROOF. We begin by showing that Σ and $\hat{\Sigma}$ are dual relative to $\gamma = \xi + f \cdot \lambda$. For this, we compute $\cup^\alpha g(x) = \tilde{E}^x \int_0^\infty e^{-\alpha t} g(\Sigma_t)dt$:

$$\cup^\alpha g(x) = (1 + \alpha f(x))^{-1}[g(x) \cdot 1_\Gamma(x) + \tilde{E}^x \int_0^\infty e^{-\alpha \tilde{B}_t} g(X_t)dt$$

$$+ \sum \sum e^{-\alpha \tilde{B}(S_{m,n})} g(X_{S_{m,n}}) \int_0^{\Delta \tilde{B}(S_{m,n})} e^{-\alpha u}du)]$$

$$= (1 + \alpha f(x))^{-1}(g(x)\cdot 1_\Gamma(x) + V^\alpha g(x) + U_D^\alpha g(x)),$$

where (V^β) is the resolvent of the subprocess (X,M) for $M_t = \prod_{0<s<t} (1 + \alpha f(X_s))^{-1}$, and D is the additive functional of (X,M) given by

$$D_t = \int_{(0,t]} M_s dA_s^d.$$

Now for $g, h \in (b\mathcal{E})^+$, $\int U^\alpha g \cdot h d\gamma$ is comprised of five terms (recall that ξ does not charge Γ):

(i) $\int V^\alpha g \cdot h\, d\xi$ (ii) $\int U_D^\alpha g \cdot h\, d\xi$

(iii) $\int V^\alpha g \cdot hf/(1 + \alpha f)d\lambda$ (iv) $\int U_D^\alpha g \cdot hf/(1 + \alpha f)d\lambda$

(v) $\int_\Gamma ghf/(1 + \alpha f)d\lambda$,

which we compare to the terms generated by $\int g \cdot \hat{U}h\, d\gamma$. First, (i) = ($\hat{\text{i}}$) by the duality of (X,M) and $(\hat{X},\hat{M})$, where

$$\hat{M}_t = \prod_{0<s<t} (1 + \alpha f(\hat{X}_s))^{-1}.$$

Next, by (2.9) of [6] the Revuz measure of D is $f/(1 + \alpha f)\cdot d\lambda$. Using (9.3) of [4], (ii) becomes

$$\int_0^\infty e^{-\alpha t}dt \int g(y)f(y)/(1 + \alpha f(y))\hat{Q}_t h(y)\lambda(dy)$$

$$= \int \hat{V}^\alpha h(y)g(y)f(y)/(1 + \alpha f(y)\lambda(dy) = (\hat{\text{iii}}).$$

Similarly (iii) = ($\hat{\text{ii}}$). Write $U_D^\alpha g(x) = \int v^\alpha(x,y)g(y)f(y)/(1 + f(y))d\lambda$, where $v^\alpha(x,y)$ is the density of $V^\alpha(x,dy)$ relative to ξ, to check that (iv) = ($\hat{\text{iv}}$). Finally, (v) = ($\hat{\text{v}}$) is obvious. This proves Σ and $\hat{\Sigma}$ are in weak duality.

The final step consists of checking that C and $\hat{C}$ are dual additive functionals. An argument like the one used in [6] shows that C and $\hat{C}$ have the same Revuz measure, and we are done.

References

1. R. M. Blumenthal and R. K. Getoor. Markov Processes and Potential Theory. Academic Press, New York 1968.

2. E. Cinlar and J. Jacod. Representation of semimartingale Markov processes in terms of Weiner processes and Poisson random measures. Seminar on Stochastic Processes 1981, 159-242. Birkhäuser, Boston, 1981.

3. E. Cinlar and H. Kaspi. Regenerative systems and Markov additive processes. Seminar on Stochastic Processes 1982, 123-147. Birkhäuser, Boston, 1983.

4. R. K. Getoor and M. J. Sharpe. Naturality, standardness and weak duality for Markov processes. Z. Wahrscheinlichkeitstheorie verw. Gebiete 67 (1984), 1-62.

5. H. Kaspi. Excursions of Markov processes: an approach via Markov additive processes. Z. Wahrscheinlichkeitstheorie verw. Gebiete 64 (1983), 251-268.

6. J. B. Mitro. A discontinuous time change for natural additive functionals which preserves duality. To appear.

7. M. J. Sharpe. Discontinuous additive functionals of dual processes. Z. Wahrscheinlichkeitstheorie verw. Gebiete 21 (1972), 81-95.

8. G. Weidenfield. Changements de temps de processus de Markov. Z. Wahrscheinlichkeitstheorie verw. Gebiete 53 (1980), 123-146.

Joanna B. Mitro
Department of Mathematical Sciences
University of Cincinnati
Cincinnati, Ohio 45221-0025

Seminar on Stochastic Processes, 1985
Birkhäuser, Boston, 1986

THE CERETELI-DAVIS SOLUTION TO THE H^1-EMBEDDING PROBLEM AND AN OPTIMAL EMBEDDING IN BROWNIAN MOTION

by

Edwin Perkins

Summary

Necessary and sufficient conditions are found on a mean-zero probability, μ, for the existence of a stopping time, T, and a Brownian motion, B, such that B_T has law μ and B_T^* is integrable. This result, due to Burgess Davis (the classical analogue was first solved by O. D. Cereteli), leads naturally to a stopping time, T, that stochastically minimizes both $\sup_{s<T} B_s$ and $-\inf_{s<T} B_s$.

1. Introduction.

Consider a mean-zero probability on the line, μ, and a one-dimensional $\{\mathscr{F}_t\}$-Brownian motion, B_t, defined on some $(\Omega, \mathscr{F}, \mathscr{F}_t, P)$ and satisfying $B_0 = 0$. An $\{\mathscr{F}_t\}$-stopping time, T, is an embedding of μ if $B(T \wedge t)$ is a uniformly integrable martingale such that $\mathscr{L}(B_t) = \mu$ ($\mathscr{L}(Z)$ denotes the law of the r.v. Z). T is an H^p-embedding if, in addition, $B_T^* \equiv \sup\{|B_s| : s < T\}$ is in L^p. The existence of an embedding is due to Skorokhod (1965). A particularly explicit one is described in Azéma-Yor (1978 a,b). If $p > 1$, Doob's strong L^p inequalities show that an H^p-embedding of μ exists iff $\int |x|^p d\mu(x) < \infty$ iff every embedding of μ is an H^p-embedding, and if $p < 1$, Doob's weak L^1 inequality shows that every embedding is an H^p-embedding. The situation for $p = 1$ is more delicate. It is well-known that there are laws μ for which some embeddings are H^1-embeddings and some are not (e.g. compare Proposition 2.1 below with Theorem 2.2 of Azéma-Yor (1978b)). A natural question, which I learned from John Walsh, is therefore:

(1.1) PROBLEM. Find necessary and sufficient conditions on μ for it to have an H^1-embedding.

Doob's LlogL inequality shows that $\int |x| \log^+|x| d\mu(x) < \infty$ ($\log^+ x = \max(\log x, 0)$) is sufficient for every embedding to be an H^1-embedding and by using optional stopping it is easy to see that it is also necessary if $\mathrm{supp}(\mu)$ ($\mathrm{supp}(\mu) \equiv$ support of μ) is bounded below or above. It is simple enough to obtain better sufficient conditions by doing some computations with one's favorite embedding, providing of course it is an embedding that allows one to compute such things as $P(\sup_{s<T} B_s > \lambda)$. This is done in the beginning of section 2, where we adopt the Skorohod embedding as our favorite embedding (it will be until section 3) and arrive at a sufficient condition due to Walsh (2.5).

Getting necessary conditions seems harder as there are a lot of embeddings to check. In fact, a complete solution to (1.1) already exists in the literature. As this seems to have escaped the notice of many probabilists, and as the history of the subject is complicated by the close connection between classical H^p-theory and probability, shown by Burkholder, Gundy and Silverstein (1971), a short historical account is in order.

Let ∂D be the unit circle in the complex plane. If f is an integrable, real-valued function on ∂D, let $\tilde{f}$ denote the conjugate function of f, and write $f \in \operatorname{Re} H^1(\partial D)$ if $\tilde{f}$ is integrable on ∂D. The space $\operatorname{Re} H^1(\partial D)$ can be considered as a subspace of the H^1-embedding stopping times and furthermore the spaces $\operatorname{Re} H^1(R^n)$, $n > 1$, are less closely, but still strongly, connected to these times (see Davis (1980) for the necessary definitions and motivation). The analogue of (1.1) for $\operatorname{Re} H^1(\partial D)$ was first answered completely by O. D. Cereteli in a series of papers (see Cereteli (1976)). He gives a condition on the distribution of an integrable function, f, on ∂D, that is necessary and sufficient for the existence of a rearrangement (of f) that belongs to $\operatorname{Re} H^1(\partial D)$. We learned of Cereteli's work from Burgess Davis, who used probability to give a different (but of course equivalent) necessary and sufficient condition in Davis (1980). (He was unaware of Cereteli's work at that time.) From a classical viewpoint the contribution of this part of Davis' paper is that the natural extension of this condition to functions on $\mathbf{R}^n$ was shown to characterize the distributions of functions in $\operatorname{Re} H^1(\mathbf{R}^n)$, whereas Cereteli's condition did not extend to this setting. For the probabilistic question (1.1) we are studying in this paper, the answer follows immediately from the probabilistic arguments in Davis (1980), as is pointed out on p. 218 of that work, and the characterization is the same as in the classical unit circle setting. However, Davis tells me that, "Any probabilist knowing Cereteli's work, as well as the results of Burkholder, Gundy and Silverstein, would have been able to answer (1.1)."

In section 2 we follow Davis' proof of necessity and show his condition is equivalent to Cereteli's condition and Walsh's sufficient condition. The main result of this section is stated as Theorem 2.7. This approach to the Cereteli-Davis theorem has the advantage of showing that if an H^1-embedding of μ exists, then the Skorokhod embedding will be such an embedding, a result we found a little surprising.

Davis' proof of necessity leads naturally to the definition of an explicit extremal embedding of μ, much in the spirit of Azema-Yor (1978a), that stochastically minimizes both $\sup_{s<T}B(s)$ and $-\inf_{s<T}B(s)$ over all embeddings T (Theorems 3.7, 3.8). This construction is carried out in section 3.

Throughout this work X denotes a random variable with the fixed law, μ, and if Y_t is a real-valued process,

$$T_Y(\lambda) = \begin{cases} \inf\{t > 0: Y_t > \lambda\} & \text{if } \lambda > 0 \\ \inf\{t > 0: Y_t < \lambda\} & \text{if } \lambda < 0 \end{cases} \qquad (\inf \emptyset = \infty).$$

We write $\lambda_n \uparrow\uparrow \lambda$ to denote that $\{\lambda_n\}$ is strictly increasing to λ.

2. The Cereteli-Davis Solution to the H^1-embedding Problem

We start by obtaining sufficient conditions for the existence of an H^1-embedding. Assume first that μ is atomless so that there is an explicit description of the Skorokhod embedding.

DEFINITION. If $\lambda > 0$, let

$$-\rho(\lambda) = \inf\{y\colon \int I(x < y \text{ or } x > \lambda)x\,d\mu(x) < 0\}.$$

It is easy to see that $\rho(\lambda) = +\infty$ iff $\mu[\lambda,\infty) = 0$ and $\rho\colon [0,\infty) \to [0,\infty]$ is non-decreasing and right-continuous. One can also show that

$$(2.1)\quad \mu[-\lambda,0] = \int I(0 < x,\ \rho(x) \le \lambda)x\rho(x)^{-1}d\mu(x).$$

If μ has a smooth, strictly positive density this is an easy calculus argument and the technical problems one faces in general are uninteresting and easily overcome. Let $R > 0$ be independent of B and have distribution function

$$P(R \le x) = \int I(y \le x)(1 + y\rho(y)^{-1})d\mu(y).$$

The right side defines a probability by (2.1). If

$$(2.2)\qquad T_s = \inf\{t\colon B_t \notin (-\rho(R),R)\},$$

then T_s is an embedding of μ. This is essentially the embedding studied in Skorokhod (1965). Although $\int x^2 d\mu(x) < \infty$ is assumed there, it is an easy exercise to check that $B(t \wedge T_s)$ is uniformly integrable without this condition.

NOTATION. $M_t = \sup_{s<t} B_s, m_t = -\inf_{s<t} B_s$, $H(\mu) = \int_0^\infty \lambda^{-1}\left|\int_{-\infty}^\infty xI(|x| > \lambda)d\mu(x)\right|d\lambda$.

PROPOSITION 2.1. Assume μ is a mean-zero, atomless probability on the line and T_s is given by (2.2).

(a) $$E(M(T_s)) = \int_0^\infty (x + \rho(x)) \log(1 + \frac{x}{\rho(x)})d\mu(x)$$

$$E(m(T_s)) = \int_0^\infty (x + \rho(x))x\rho(x)^{-1}\log(1 + \frac{\rho(x)}{x})d\mu(x)$$

(b) $$E(M(T_s) + m(T_s)) < 2(\int_{-\infty}^\infty |x|d\mu(x) + \int_0^\infty x\left|\log \frac{x}{\rho(x)}\right|d\mu(x))$$

$$= 2(\int_0^\infty |x|d\mu(x) + H(\mu)).$$

PROOF.

(a) $$P(M(T_s) > \lambda) = \int P(M(T_s) > \lambda|R = x)(1 + \frac{x}{\rho(x)})d\mu(x)$$

$$= \int_\lambda^\infty \frac{\rho(x)}{\lambda + \rho(x)} (1 + \frac{x}{\rho(x)})d\mu(x)$$

$$E(M(T_s)) = \int_0^\infty \int_\lambda^\infty \frac{x + \rho(x)}{\lambda + \rho(x)}d\mu(x)d\lambda$$

$$= \int_0^\infty (x + \rho(x))\log(1 + \frac{x}{\rho(x)})d\mu(x).$$

A similar argument gives the required expression for $E(m(T_s))$.

(b) Use the inequalities $\log(1 + y) < y$ and $\log(1 + y) < 1 + |\log y|$ for all $y > 0$, to see that

$$\int_0^{\infty} (x + \rho(x)) \log(1 + \frac{x}{\rho(x)}) d\mu(x)$$

$$+ \int_0^{\infty} (x + \rho(x)) \frac{x}{\rho(x)} \log(1 + \frac{\rho(x)}{x}) d\mu(x)$$

$$\leq \int_0^{\infty} x(1 + \left|\log \frac{x}{\rho(x)}\right|) + x \, d\mu(x)$$

$$+ \int_0^{\infty} x + x(1 + \left|\log(\frac{\rho(x)}{x})\right|) d\mu(x)$$

$$= 2(\int_{-\infty}^{\infty} |x| d\mu(x) + \int_0^{\infty} x \left|\log \frac{x}{\rho(x)}\right| d\mu(x)),$$

and hence obtain the first inequality in (b).

Let $\lambda_n \uparrow\uparrow \lambda$ and take limits in

$$\int x \, I(x < -\rho(\lambda_n) \text{ or } x > \lambda_n) d\mu(x) = 0$$

to see that

$$\int x \, I(x < -\rho(\lambda-) \text{ or } x > \lambda) d\mu(x) = 0.$$

As the same equation holds with $\rho(\lambda)$ in place of $\rho(\lambda-)$, it must be that

(2.3) $\qquad \mu[-\rho(\lambda), -\rho(\lambda-)] = 0 \quad$ for each $\lambda > 0$.

In particular, if $\rho^{-1}(x)$ denotes the right-continuous inverse of ρ, then $-\lambda < -\rho(\rho^{-1}(\lambda)-)$ and so $\mu[-\rho(\rho^{-1}(\lambda)), -\lambda] = 0$. It follows that

$$\int x \, I(x < -\lambda \text{ or } x > \rho^{-1}(\lambda)) d\mu(x)$$

$$= \int x\, I(x < -\rho(\rho^{-1}(\lambda)) \text{ or } x > \rho^{-1}(\lambda))d\mu(x)$$

$$= 0,$$

and therefore

$$(2.4) \qquad \left|\int xI(x < -\lambda \text{ or } x > \lambda)d\mu(x)\right|$$

$$= \int xI(\lambda \wedge \rho^{-1}(\lambda) < x < \lambda \vee \rho^{-1}(\lambda))d\mu(x).$$

An argument similar to that used to show (2.3) gives us $\mu[r,s] = 0$ whenever $r < s$ satisfy $\rho(r) = \rho(s)$. This implies that μ does not charge the "flat spots" of ρ and hence that $\rho^{-1}(\rho(x)) = x$ for μ-a.a. $x > 0$.
This gives us

$$\rho^{-1}(\lambda) < x \iff \lambda < \rho(x) \qquad \text{for } \mu\text{-a.a.} \quad x > 0$$

and hence, by (2.4),

$$\left|\int x\, I(x < -\lambda \text{ or } x > \lambda)d\mu(x)\right|$$

$$= \int_0^\infty x\, I(x \wedge \rho(x) < \lambda < \rho(x) \vee x)d\mu(x).$$

It follows that

$$H(\mu) = \int_0^\infty \int_{x\wedge\rho(x)}^{x\vee\rho(x)} \lambda^{-1}d\lambda\; x\, d\mu(x) = \int_0^\infty x\left|\log \frac{x}{\rho(x)}\right|d\mu(x)$$

and the proof is complete. □

As an immediate corollary we see that, when μ is atomless, either of the two equivalent conditions

$$(2.5) \qquad \int_0^\infty x\left|\log \frac{x}{\rho(x)}\right| d\mu(x) < \infty$$

or

$$(2.6) \qquad H(\mu) < \infty$$

is sufficient for the existence of an H^1-embedding, namely T_s. These conditions are symmetry conditions on the tails of μ. Both conditions hold if μ is symmetric (the integrals are zero) and are equivalent to $\int|x|\log^+|x|d\mu(x) < \infty$ if supp(μ) is bounded above or below. (2.6) appears in Cereteli (1976), and also in Vallois (1982). (2.5) was shown by Walsh (private communication) to be necessary and sufficient for $E(B^*(T_s)) < \infty$ and led him to make the

(2.7) CONJECTURE (2.5) is necessary and sufficient for the existence of an H^1-embedding of the atomless measure μ.

The necessity of (2.7) is not at all obvious, since T_s is in no way an "optimal embedding".

Our immediate task is to extend these results to the case when μ may have atoms. It will be easier to work with (2.6) than (2.5). Let α_n denote the uniform law on $[-\frac{1}{n},\frac{1}{n}]$, $\mu_n = \alpha_n * \mu$ (* denotes convolution and $T_{s,n}$ denote the Skorokhod embedding of μ_n. Then there are random variables $(U_n, V_n) \in (-\infty, 0] \times [0, \infty)$, independent of B such that

$$T_{s,n} = \inf\{t > 0: B_t \notin (U_n, V_n)\}.$$

By changing the underlying probability space we may assume there is a subsequence such that $(U_{n_k}, V_{n_k}) \overset{a.s.}{\to} (U,V) \in [-\infty, 0] \times [0, \infty]$, where $\{U_{n_k}, V_{n_k} : k \in \mathbf{N}\}$ is independent of the Brownian motion B (Skorokhod (1965, Ch. 1.6)). It follows that

$$T_{s,n_k} \overset{a.s.}{\to} T_s \equiv \inf\{t > 0: B_t \notin (U,V)\} < \infty.$$

Note also that

$$P(T_{s,n} > K) < P(T_{s,n} > K,\ B^*_{T_{s,n}} < M) + P(B^*_{T_{s,n}} > M)$$

$$< P(B^*_K < M) + M^{-1}E(|B_{T_{s,n}}|)$$

$$< P(B^*_1 < M\, K^{-1/2}) + M^{-1}(\int |x| d\mu(x) + n^{-1}).$$

The last line may be made arbitrarily small, uniformly in n, by first choosing M and then K large enough. Therefore $T_s < \infty$ a.s. and so $B(T_{s,n_k}) \overset{a.s.}{\to} B(T_s)$. This implies $\mathscr{L}(B(T_s)) = \mu$. Moreover we have

$$\lim_{k\to\infty} E|B(T_{s,n_k})| = \lim_{k\to\infty} \int |x| d\mu_{n_k}(x) = E(|B(T_s)|).$$

This clearly shows $B(T_s \wedge t)$ is uniformly integrable and hence T_s is an embedding of μ, which we call the Skorokhod embedding of μ (although, given this nebulous procedure, "the" may be rather strong language).

THEOREM 2.2. Let μ be a mean-zero probability on the line and let T_s be defined as above. Then

$$E(M_{T_s} + m_{T_s}) \le 2(\int |x| d\mu(x) + H(\mu)),$$

and in particular $H(\mu) < \infty$ implies T_s is an H^1-embedding of μ.

PROOF. Recall X denotes a r.v. with law μ. Let U_n be independent of X and have law α_n, and let $X_n = X + U_n$. Then

$$\left|\int_1^\infty \lambda^{-1}(|E(X_n I(|X_n| > \lambda))| - |E(X\ I(|X| > \lambda))|)d\lambda\right|$$

$$\le \int_1^\infty \lambda^{-1}|E(X_n I(|X_n| > \lambda)) - E(X\ I(|X| > \lambda))|d\lambda$$

$$\le \int_1^\infty \lambda^{-1}(n^{-1}P(|X_n| > \lambda) + |E(X(I(|X| > \lambda)$$

$$- I(|X_n| > \lambda)))|)d\lambda$$

$$\le n^{-1}E(\log^+|X_n|)$$

$$+ \int_1^\infty \lambda^{-1}E(|X|I(|X - \lambda| < n^{-1} \text{ or } |X + \lambda| < n^{-1}))d\lambda$$

$$\le n^{-1}(E(\log^+|X_n|) + 4\ E|X|) \to 0 \quad \text{as } n \to \infty.$$

Note also that

$$\lim_{n\to\infty} \int_0^1 \lambda^{-1}|E(X_n I(|X_n| > \lambda))|d\lambda = \int_0^1 \lambda^{-1}|E(X\ I(|X| > \lambda))|d\lambda$$

because the integrands converge for Lebesgue-a.a. λ and are bounded by one (recall $E(X_n) = 0$). We have shown that

$$\lim_{n\to\infty} H(\mu_n) = H(\mu). \tag{2.8}$$

Let $\{n_k\}$ be the subsequence used to construct T_s. Then

$$E(M(T_s) + m(T_s)) \leq \liminf_{k\to\infty} E(M(T_{s,n_k}) + m(T_{s,n_k}))$$
(Fatou's lemma)

$$\leq \liminf_{k\to\infty} 2(\int_{-\infty}^{\infty} |x|d\mu_{n_k}(x) + H(\mu_{n_k}))\ \text{(Proposition 2.1(b))}$$

$$= 2(\int_{-\infty}^{\infty} |x|d\mu(x) + H(\mu))\ \text{(by (2.8))}. \square$$

To find necessary conditions for the existence of an H^1-embedding introduce the

NOTATION $-\alpha = -\alpha^{\mu} = \inf \operatorname{supp}(\mu)$, $\beta = \beta^{\mu} = \sup \operatorname{supp}(\mu)$

$$-\gamma_+(\lambda) = -\gamma_+^{\mu}(\lambda)$$

$$= \begin{cases} \sup\{y: E(X|X<y \text{ or } X>\lambda) > \lambda\} & \text{if } P(X>\lambda) > 0 \\ -\alpha & \text{otherwise} \end{cases}$$

$$= \sup\{y: \int (x - \lambda)I(x < y \text{ or } x > \lambda)d\mu(x) > 0\}$$

$$(\sup \emptyset = -\alpha,\ \lambda > 0)$$

$$\gamma_-(\lambda) = \gamma_-^{\mu}(\lambda)$$

$$= \begin{cases} \inf\{y\colon E(X|X < -\lambda \text{ or } X > y) < -\lambda\} & \text{if } P(X < -\lambda) > 0 \\ \beta & \text{otherwise} \end{cases}$$

$$= \inf\{y\colon \int (x + \lambda) I(x < -\lambda \text{ or } x > y) < 0\} \quad (\inf \emptyset = \beta, \lambda > 0)$$

$$\psi(y) = \begin{cases} \int x\, I(x > y) d\mu(x)/\mu[y,\infty) & \text{if } \mu[y,\infty) > 0 \\ y & \text{if } \mu[y,\infty) = 0 \end{cases}$$

$$\phi(\lambda) = \inf\{y\colon \psi(y) > \lambda\} \quad (\lambda > 0)$$

$$\tilde{\mu}(-\infty,x] = \mu[-x,\infty).$$

Hence ϕ is the increasing left-continuous inverse of the increasing, left-continuous barycentre function ψ (see Azema-Yor (1978a)). $\gamma_{\pm}$ are increasing left-continuous functions from $[0,\infty)$ to $[0,\infty)$. These and other properties of $\gamma_{\pm}$ will be discussed in the next section (Lemma 3.2). For now, we will need the following results, which follow easily from the definitions:

(2.9) $$\int (x - \lambda) I(x > \phi(\lambda)) d\mu(x) < 0$$

$$< \int (x - \lambda) I(x > \phi(\lambda)) d\mu(x)$$

(2.10) $$\int (x - \lambda) I(x < -\gamma_+(\lambda) \text{ or } x > \lambda) d\mu(x) < 0$$

$$< \int (x - \lambda) I(x < -\gamma_+(\lambda) \text{ or } x > \lambda d\mu(x)$$

(2.11) $$\int (x + \lambda) I(x < -\lambda \text{ or } x > \gamma_-(\lambda)) d\mu(x) < 0$$

$$< \int (x + \lambda) I(x < -\lambda \text{ or } x > \gamma_-(\lambda)) d\mu(x)$$

(2.12) $\gamma_-^{\mu} = \gamma_+^{\tilde{\mu}}$.

NOTATION. If $\lambda > 0$, let

$$p(\lambda) = \begin{cases} \int (x - \lambda) I(x > \phi(\lambda)) d\mu(x) \, (\lambda - \phi(\lambda))^{-1} & \text{if } \phi(\lambda) < \lambda \\ \mu(\{\phi(\lambda)\}) & \text{if } \phi(\lambda) = \lambda \end{cases}$$

$$q_+(\lambda) = \int (x - \lambda) I(x < -\gamma_+(\lambda) \text{ or } x > \lambda) d\mu(x) (\gamma_+(\lambda) + \lambda)^{-1}$$

$$q_-(\lambda) = \int (x + \lambda) I(x < -\lambda \text{ or } x > \gamma_-(\lambda)) d\mu(x) (\gamma_-(\lambda) + \lambda)^{-1}$$

$$\mu^*(\lambda) = \mu(\phi(\lambda), \infty) + p(\lambda)$$

$$\mu_+(\lambda) = \mu(-\infty, -\gamma_+(\lambda)) + q_+(\lambda) + \mu[\lambda, \infty)$$

$$\mu_-(\lambda) = \mu(\gamma_-(\lambda), \infty) + q_-(\lambda) + \mu(-\infty, -\lambda].$$

LEMMA 2.3. (2.13) $0 < p(\lambda) < \mu(\{\phi(\lambda)\})$

(2.14) $0 < q_+(\lambda) < \mu(\{-\gamma_+(\lambda)\})$

(2.15) $0 < q_-(\lambda) < \mu(\{\gamma_-(\lambda)\})$

(2.16) $q_-^{\mu}(\lambda) = q_+^{\tilde{\mu}}(\lambda)$, $\mu_- = \tilde{\mu}_+$

PROOF. (2.9) implies that

$$0 < \int (x - \lambda) I(x > \phi(\lambda)) d\mu(x) < (\lambda - \phi(\lambda)) \mu(\{\phi(\lambda)\}).$$

Divide the above by $\lambda - \phi(\lambda)$ to obtain (2.13). Similarly one can use (2.10) and (2.11) to prove (2.14) and (2.15). (2.16) is an easy consequence of (2.12). □

The key idea in the derivation of necessary conditions for the existence of an H^1-embedding is

LEMMA 2.4 (Davis' Law of the Lever). *Assume* $\mathscr{L}(X) = \mu$, $\lambda > 0$ *and* A *is a measurable set such that* $\{X \ge \lambda\} \subset A$.

(a) (i) *If* $\int_A X - \lambda \, dP \ge 0$, *then*

$$P(A) \le \mu^*(\lambda). \tag{2.17}$$

(ii) *If, in addition, equality holds in (2.17), then*

$$\{X > \phi(\lambda)\} \subset A \subset \{X \ge \phi(\lambda)\} \text{ a.s.} \tag{2.18}$$

(iii) *Conversely if (2.18) holds and* $\int_A X - \lambda \, dP = 0$ *then* $P(A) = \mu^*(\lambda)$.

(b) (i) *If* $\int_A X - \lambda \, dP \le 0$, *then*

$$P(A) \ge \mu_+(\lambda). \tag{2.19}$$

(ii) *If, in addition, equality holds in (2.19), then*

$$\{X \le -\gamma_+(\lambda) \text{ or } X \ge \lambda\} \subset A \subset \{X \le -\gamma_+(\lambda) \text{ or } X \ge \lambda\} \text{ a.s.} \tag{2.20}$$

(iii) Conversely if (2.20) holds and $\int_A X - \lambda dP = 0$, then $P(A) = \mu_+(\lambda)$.

REMARK. (2.17) was observed in Blackwell-Dubins (1963). The idea of (b) appears in Davis (1980, p. 215, 1982, p. 157).

These results are intuitively obvious. Sand is distributed along a see-saw according to μ. The fulcrum is at λ and sand is initially added to the right of λ. (a) says that if we want to add the maximum amount of sand without tipping the see-saw to the left, we should add it as close as possible to the fulcrum. (b) says that if we want to add the minimum amount of sand needed to tip the see-saw to the left or at least put it in equilibrium, we should add it as far from the fulcrum as possible. Although a proof is clearly not needed, we include a derivation of (b) because of its importance in what follows.

PROOF OF (b). The definition of $q_+(\lambda)$ gives us

$$\int_A X - \lambda \, dP \leq 0 = \int I(X < -\gamma_+(\lambda) \text{ or } X > \lambda)(X - \lambda)dP$$
$$- (\gamma_+(\lambda) + \lambda)q_+(\lambda)$$

and therefore

(2.21) $\int I(A, -\gamma_+(\lambda) \leq X \leq \lambda)(X - \lambda)dP$

$$\leq \int I(A^c, X < -\gamma_+(\lambda))(X - \lambda)dP - (\gamma_+(\lambda) + \lambda)q_+(\lambda).$$

This implies that

$$(2.22)\quad (-\gamma_+(\lambda) - \lambda)P(A, -\gamma_+(\lambda) \leq X < \lambda) \leq \text{LHS of } (2.21) \leq$$

$$\text{RHS of } (2.21) \leq (-\gamma_+(\lambda) - \lambda)[P(A^c, X < -\gamma_+(\lambda)) + q_+(\lambda)].$$

If $\gamma_+(\lambda) = \infty$, (2.19) is trivial. Assuming $\gamma_+(\lambda) < \infty$, we may divide the above by $-\gamma_+(\lambda) - \lambda$ and then add $P(A, X < -\gamma_+(\lambda)) + P(X \geq \lambda)$ to both sides to complete the proof of (2.19).

If $P(A) = \mu_+(\lambda)$, then reversing the final steps in the above argument, we see that the extreme left and right sides of (2.22) are equal. This means that

$$\int I(A, -\gamma_+(\lambda) \leq X < \lambda)(-\gamma_+(\lambda) - \lambda)dP$$

$$= \int I(A, -\gamma_+(\lambda) \leq X < \lambda)(X - \lambda)dP$$

$$= \int I(A^c, X < -\gamma_+(\lambda))(X - \lambda)dP - (\gamma_+(\lambda) + \lambda)q_+(\lambda)$$

$$= \int I(A^c, X < -\gamma_+(\lambda))(-\gamma_+(\lambda) - \lambda)dP - (\gamma_+(\lambda) + \lambda)q_+(\lambda).$$

The last equality implies that $P(A^c, X < -\gamma_+(\lambda)) = 0$ and hence the first inclusion in (2.20) holds. The first equality shows that $P(A, -\gamma_+(\lambda) < X < \lambda) = 0$ and hence the second inclusion in (2.20) holds.

Finally note that, under the hypotheses of (iii), if

$$A_\lambda \equiv A - \{X < -\gamma_+(\lambda) \text{ or } X > \lambda\}$$

then $A_\lambda \subset \{X = -\gamma_+(\lambda)\}$ and so

$$0 = \int_A (X - \lambda)dP$$

$$= \int I(X < -\gamma_+(\lambda) \text{ or } X > \lambda)dP - (\gamma_+(\lambda) + \lambda)P(A_\lambda).$$

Solve for $P(A_\lambda)$ to get $P(A_\lambda) = q_+(\lambda)$ and therefore $P(A) = \mu_+(\lambda)$. □

THEOREM 2.5. <u>Assume</u> $\{X_t: t > 0\}$ <u>is a uniformly integrable (right-continuous) martingale such that</u> $\mathscr{L}(X_\infty) = \mu$.

(a) (Blackwell-Dubins (1963)). <u>For all</u> $\lambda > 0$,

(2.23) $$P(\sup_t X_t > \lambda) < \mu^*(\lambda).$$

<u>If equality holds in (2.23) then</u>

(2.24) $$\{X_\infty > \phi(\lambda)\} \subset \{\sup_t X_t > \lambda\} \subset \{X_\infty > \phi(\lambda)\} \text{ a.s.}$$

<u>Conversely if</u> X_t <u>is a.s. continuous,</u> $X_0 = 0$ <u>and (2.24) holds, then equality holds in (2.23)</u>.

(b) (Davis) <u>Assume, in addition that</u> X_t <u>is a.s. continuous and</u> $X_0 = 0$.

<u>Then for all</u> $\lambda > 0$,

(2.25) $$P(\sup_t X_t > \lambda) > \mu_+(\lambda)$$

(2.26) $$P(-\inf X_t > \lambda) > \mu_-(\lambda).$$

<u>Equality holds in (2.25) (respectively, (2.26)) iff</u>

$$(2.27)\quad \{X_\infty < -\gamma_+(\lambda) \text{ or } X_\infty > \lambda\} \subset \{\sup_t X_t > \lambda\}$$

$$\subset \{X_\infty < -\gamma_+(\lambda) \text{ or } X_\infty > \lambda\} \text{ a.s.}$$

(<u>respectively,</u>

$$(2.28)\quad \{X_\infty < -\lambda \text{ or } X_\infty > \gamma_-(\lambda)\} \subset \{-\inf_t X_t > \lambda\}$$

$$\subset \{X_\infty < -\lambda \text{ or } X_\infty > \gamma_-(\lambda)\} \text{ a.s.)}$$

REMARK. It is not hard to show that the right side of (2.23) equals $\bar{\mu}[\lambda,\infty)$ where $\bar{\mu}$ is the distribution of the Hardy-Littlewood maximal function associated with μ (see e.g. Dubins-Gilat (1978)). Thus (2.23) really is Theorem 3(a) of Blackwell-Dubins (1963) (see also Theorem 1 of Dubins-Gilat (1978)).

PROOF. (b) The optional stopping theorem shows that for $\lambda > 0$,

$$\int I(X(T_X(\lambda)) > \lambda)(X_\infty - \lambda)dP$$

$$= \int I(X(T_X(\lambda)) > \lambda)(X(T_X(\lambda)) - \lambda)dP = 0.$$

Apply Lemma 2.4(b) with $A = \{X(T_X(\lambda)) > \lambda\}$ and $X = X_\infty$ to obtain (2.25), and the equivalence between (2.27) and equality holding (2.25). The rest of (b) is obtained by replacing X with -X and μ with $\tilde{\mu}$ (use (2.12) and (2.16) here).

(a) Use Lemma 2.4(a) as above. In this case the possibility of jumps as well as X_0 exceeding λ means that

$$\int I(X(T_X(\lambda)) > \lambda)(X(T_X(\lambda)) - \lambda)dP > 0.$$

Therefore continuity of X and $X_0 = 0$ is needed for the last statement in (a). □

By integrating out (2.25) and (2.26) we see that a necessary condition for the existence of an H^1-embedding of μ is

$$\int_0^\infty \mu_+(\lambda) + \mu_-(\lambda)d\lambda < \infty.$$

It remains to show that this is equivalent to our earlier sufficient condition, $H(\mu) < \infty$.

LEMMA 2.6.

$$(2.29) \qquad H(\mu) < 2\int_0^\infty \mu_+(\lambda) + \mu_-(\lambda)d\lambda$$

$$(2.30a) \qquad \int_0^\infty \mu_+(\lambda)d\lambda < \int_{-\infty}^\infty |x|d\mu(x) + H(\mu)$$

$$(2.30b) \qquad \int_0^\infty \mu_-(\lambda)d\lambda < \int_{-\infty}^\infty |x|d\mu(x) + H(\mu)$$

PROOF. Fix $\lambda > 0$ and note that

$$(2.31) \quad \int I(x < -\gamma_+(\lambda) \text{ or } x > \lambda)(x - \lambda)d\mu(x)$$

$$- (\lambda + \gamma_+(\lambda))q_+(\lambda) = 0.$$

case 1. $\lambda > \gamma_+(\lambda)$

$$\int_{-\infty}^{\infty} xI(|x| > \lambda)d\mu(x) = \int_{-\infty}^{\infty} xI(|x| > \lambda)d\mu(x) -(2.31)$$

$$= -\int_{-\infty}^{\infty} xI(-\lambda < x < -\gamma_+(\lambda))d\mu(x) + \gamma_+(\lambda)q_+(\lambda) + \lambda\mu_+(\lambda)$$

$$\Rightarrow \mu_+(\lambda) < \lambda^{-1}\left|\int_{-\infty}^{\infty} xI(|x| > \lambda)d\mu(x)\right| \text{ (by (2.14))}$$

$$(2.32) \quad < \int_{-\infty}^{\infty} \frac{-x}{\lambda} I(-\lambda < x < -\gamma_+(\lambda))d\mu(x)$$

$$+ \frac{\gamma_+(\lambda)}{\lambda}q_+(\lambda) + \mu_+(\lambda) < 2\mu_+(\lambda).$$

case 2. $\lambda > \gamma_-(\lambda)$.

Replace μ with $\tilde{\mu}$ in the above to get

$$(2.33) \qquad \mu_-(\lambda) < \lambda^{-1}\left|\int_{-\infty}^{\infty} xI(|x| > \lambda)d\mu(x)\right| < 2\mu_-(\lambda).$$

case 3. $\lambda < \gamma_+(\lambda)$ and $\lambda < \gamma_-(\lambda)$

$$\int_{-\infty}^{\infty} xI(|x| > \lambda)d\mu(x) = \int_{-\infty}^{\infty} xI(|x| > \lambda)d\mu(x) - (2.31)$$

$$= \int_{-\infty}^{\infty} xI(-\gamma_+(\lambda) < x < -\lambda)d\mu(x) + \gamma_+(\lambda)q_+(\lambda) + \lambda\mu_+(\lambda)$$

$$< \int_{-\infty}^{\infty} xI(-\gamma_+(\lambda) < x < -\lambda)d\mu(x) + \lambda\mu_+(\lambda) \text{ (by (2.14))}$$

$$< \lambda\mu_+(\lambda).$$

By symmetry we may conclude that

$$-\lambda\mu_-(\lambda) < \int xI(|x| > \lambda)d\mu(x) < \lambda\mu_+(\lambda)$$

(2.34) $$\lambda^{-1}|\int xI(|x| > \lambda)d\mu(x)| < \mu_+ \lambda) + \mu_-(\lambda).$$

(2.29) follows by using the upperbounds on $\lambda^{-1}|\int xI(|x| > \lambda)d\mu(x)|$ in (2.32), (2.33), (2.34) and then integrating out λ.

For (2.30a) note that if $\lambda < \gamma_+(\lambda)$ then (2.14) shows that

(2.35) $$\mu_+(\lambda) < \mu(\{|x| > \lambda\})$$

This, together with the first inequality in (2.32) gives (2.30a) upon integrating. Replace μ with $\tilde{\mu}$ to get (2.30b) from (2.30a). □

It is now an easy matter to prove the main result of this section. Recall the definitions of $H(\mu)$ and $\mu_\pm(\lambda)$ given prior to Proposition 2.1 and Lemma 2.3, respectively.

THEOREM 2.7. *Let* μ *be a mean-zero probability on the line and* T_s *be the Skorokhod embedding of* μ. *The following are equivalent:*

(a) *There exists an* H^1*-embedding of* μ.

(b) T_s *is an* H^1*-embedding of* μ.

(c) $H(\mu) < \infty$

(d) $\int_0^\infty \mu_+(\lambda) + \mu_-(\lambda)d\lambda < \infty$.

PROOF. (d) <=> (c) Lemma 2.6

(c) => (b) Theorem 2.2

(b) => (a) obvious

(a) => (d) Theorem 2.4 (b). □

REMARKS. 1. In particular the equivalence of (2.5) and (d) in the atomless case (Prop. 2.1(b)) shows that Walsh's conjecture (2.7) is true.

2. It is a little surprising that if an H^1-embedding of μ exists then T_s must be an H^1-embedding. Clearly there must be other embeddings with this property. Vallois (1982) describes an interesting embedding, T_V, that uses local time, and shows if $\mu\{0\} = 0$, it is an H^1-embedding iff(c) holds in the above (Vallois (1982), Prop. 4.23)). The filling scheme, T_C, has been studied extensively (see e.g. Rost (1971), Baxter [3]) and is known to minimize $E(\sqrt{T})$ over all embeddings of μ (see P. Chacon (1985)). Davis' inequality shows that if an H^1-embedding exists then T_C is such an embedding. Indeed, this suggested a direct method of attack on the original problem, namely find NASC on μ for $E(B^*_{T_C}) < \infty$. Unfortunately, the filling scheme does not seem to lend itself to such explicit calculations.

3. An Optimal Embedding

How should one define an embedding of μ, T, that minimizes B^*_T, M_T or m_T? Theorem 2.5 tells us how one might hope to define such a T. To illustrate the idea let us first try to maximize M_T. According to Theorem 2.5(a), $P(M_T \geq \lambda) \leq \mu^*(\lambda)$, and, if equality holds, T must satisfy

$$\{B_T > \phi(\lambda)\} \subset \{M_T \geq \lambda\} \subset \{B_T \geq \phi(\lambda)\}.$$

The left-continuity of ψ now implies

$$\psi(B_T) > \lambda \Rightarrow B_T > \phi(\lambda) \Rightarrow M_T \geq \lambda.$$

Let $\lambda \uparrow \psi(B_T)$ to conclude that $M_T \geq \psi(B_T)$. This suggests the

DEFINITION. $T_a = \inf\{t > 0: M_t \geq \psi(B_t)\}$ $(\inf \phi = \infty)$.

Only an optimist would expect T_a to be an embedding of μ. In fact it is precisely the embedding studied by Azéma and Yor (1978a,b). The point of this digression is that Theorem 2.5 provides a natural route to their stopping time. Moreover, it is now easy to show that T_a stochastically maximizes M_T over all embeddings T, as was observed in Azéma -Yor (1978b).

THEOREM 3.1.

(3.1) $P(M(T_a) \geq \lambda) = \mu^*(\lambda)$ <u>for all</u> $\lambda \geq 0$.

<u>If</u> $\{X_t: t \geq 0\}$ <u>is a uniformly integrable (right-continuous) martingale such that</u> $\mathscr{L}(X_\infty) = \mu$, <u>then</u>

$$P(\sup_t X_t \geq \lambda) \leq P(M(T_a) \geq \lambda) \quad \underline{\text{for all}}\ \lambda \geq 0.$$

PROOF. We first show that

$$(3.2)\qquad \Psi(B(T_a)) \leq M(T_a).$$

If (3.2) fails there must at least exist a sequence $t_n \downarrow\downarrow T_a$ such that $\psi(B(t_n)) < M(t_n)$. Choose $u_n \in (T_a, t_n]$ such that $B(u_n) \leq \min(B(T_a), B(t_n))$ (if $B(t_n) \leq B(T_a)$, let $u_n = t_n$). Then $\psi(B_{u_n}) < M(t_n)$ and $B(u_n) \leq B(T_a)$, so, letting $n \to \infty$ in the first inequality, we get (3.2) by the left-continuity of ψ. It follows that for each $\lambda > 0$,

$$(3.3)\qquad \{B(T_a) > \phi(\lambda)\} \subset \{M(T_a) \geq \lambda\}.$$

The definition of T_a allows one to conclude $\{B(T_a) < \phi(\lambda)\} \subset \{M(T_a) < \lambda\}$ and hence for each $\lambda > 0$,

$$(3.4)\qquad \{M(T_a) \geq \lambda\} \subset \{B(T_a) \geq \phi(\lambda)\}.$$

(3.3) and (3.4) allow us to apply Theorem 2.5 (a) with $X = B(T_a)$ and $A = \{M(T_a) \geq \lambda\}$ and conclude that (3.1) holds. The rest of the result is then immediate from the Blackwell-Dubins theorem (Theorem 2.5(a)).□

To stochastically minimize M_T, use Theorem 2.5(b) to show that if $P(M_T \geq \lambda) = \mu_+(\lambda)$, then T must satisfy

$$(3.5)\qquad \{B_T < -\gamma_+(\lambda) \text{ or } B_T > \lambda\} \subset \{M_T \geq \lambda\}$$

$$\subset \{B_T \leq -\gamma_+(\lambda) \text{ or } B_T \geq \lambda\}.$$

Using the latter inclusion and letting $\lambda = M_T$, we see that

(3.6) if $B_T < 0$, then $B_T < -\gamma_+(M_T)$.

To simultaneously minimize m_T, we see in the same way that T should also satisfy

(3.7) if $B_T > 0$, then $B_T > \gamma_-(m_T)$.

(3.6) and (3.7) together suggest the

DEFINITION. $T_d = \inf\{t > 0: B_t \notin (-\gamma_+(M_t), \gamma_-(m_t))\}$ $(\inf \emptyset = \infty)$.

There is a slight problem with this definition. If $\mu_\alpha = \alpha\mu + (1 - \alpha)\delta_0 (0 < \alpha < 1)$, then it is easy to see that $\gamma_\pm^{\mu_\alpha} = \gamma_\pm^\mu$ and hence T_d would be the same for all of these laws. To handle atoms at zero we may, and shall, assume our probability space is rich enough to support a r.v., U, uniformly distributed on [0,1] and independent of B, and make the

DEFINITION. $T_b = \begin{cases} T_d & \text{if } U > \mu(\{0\}) \\ 0 & \text{if } U < \mu(\{0\}) \end{cases}$.

We sometimes write T_b^μ or T_d^μ to denote the dependence on μ.

The optimality properties of T_b are fairly easy to show, once one knows that T_b is an embedding of μ. For this we need some further properties of $\gamma_\pm$.

LEMMA 3.2. (a) γ_+ _and_ γ_- _are non-decreasing, left-continuous functions from_ $[0,\infty)$ _to_ $[0,\alpha]$ _and_ $[0,\beta]$, *respectively.*

(b) $\lambda < \beta \Rightarrow \gamma_+(\lambda) < \infty$, $\lambda > \beta \Rightarrow \gamma_+(\lambda) = \alpha$

$\lambda < \alpha \Rightarrow \gamma_-(\lambda) < \infty$, $\lambda > \alpha \Rightarrow \gamma_-(\lambda) = \beta$

(c) $\lambda_{\pm}(\lambda) > 0$ *if* $\lambda > 0$.

(d) *If* $a, b > 0$, $a + b > 0$, $\gamma_+(b) < a$, *and* $\gamma_-(a) < b$, *then* $\mu([-a,b]^c) = 0$.

PROOF. By replacing μ with $\tilde{\mu}$, it suffices to consider γ_+.

(a) It is clear from the definition that γ_+ is non-decreasing and takes values in $[0,\alpha]$. If $\lambda_n \uparrow \lambda$ and $y > -\gamma_+(\lambda)$, then

$$0 > \int (x - \lambda)I(x < y \text{ or } x > \lambda)d\mu(x)$$

$$= \lim_{n\to\infty} \int (x - \lambda_n)I(x < y \text{ or } x > \lambda_n)d\mu(x),$$

and so for large enough n we have $y > -\gamma_+(\lambda_n)$. This shows that $-\gamma_+(\lambda) > \lim_{n\to\infty} -\gamma_+(\lambda_n)$. As the opposite inequality is obvious by monotonicity, we see that γ_+ is left continuous.

(b) If $\lambda < \beta$, then

$$\lim_{y\to-\infty} \int (x - \lambda)I(x < y \text{ or } x > \lambda)d\mu(x)$$
$$= \int (x - \lambda)I(x > \lambda)d\mu(x) > 0$$

and hence $-\gamma_+(\lambda) > -\infty$. If $\lambda > \beta$ and $y > -\alpha$ then

$$\int (x - \lambda)I(x < y \text{ or } x > \lambda)d\mu(x) = \int (x - \lambda)I(x < y)d\mu(x) < 0$$

and so $-\gamma_+(\lambda) = -\alpha$.

(c) If $\lambda > 0$, then

$$\lim_{\varepsilon \to 0+} \int (x - \lambda) I(x < -\varepsilon \text{ or } x > \lambda) d\mu(x)$$

$$= \int (x - \lambda) I(x < 0 \text{ or } x > \lambda) d\mu(x) \leq -\lambda\mu(-\infty, 0) < 0.$$

Thus the integral on the left is negative for ε small enough and for such an ε, $\gamma_+(\lambda) \geq \varepsilon > 0$.

(d) $\gamma_+(b) \leq a$ and $\gamma_-(a) \leq b$, together with (2.10) and (2.11) give

$$\int I(x < -a \text{ or } x > b)(x + a) d\mu(x) \leq 0$$

$$\int I(x < -a \text{ or } x > b)(x - b) d\mu(x) \geq 0.$$

Subtracting, we get

$$(a + b)\mu([-a,b]^c) \leq 0,$$

and hence the result. □

NOTATION. Let $\sigma_\pm(\gamma) = \sigma^\mu_\pm(\gamma)$ denote the left-continuous inverse of $\gamma_\pm$, i.e., $\sigma_\pm(\gamma) = \inf\{\lambda \geq 0 : \gamma_\pm(\lambda) \geq \gamma\}$ ($\inf \phi = +\infty$).

LEMMA 3.3. (a) $\gamma < \alpha \Rightarrow \sigma_+(\gamma) < \infty$,

$\gamma < \beta \Rightarrow \sigma_-(\gamma) < \infty$

(b) $\sigma_\pm(0+) = 0$

(c)

$$(3.8)\quad \int I(x < -\gamma \text{ or } x > \sigma_+(\gamma))(x - \sigma_+(\gamma))d\mu(x) = 0$$

$$(0 \cdot (-\infty) = 0)$$

$$(3.9)\quad \int I(x < -\sigma_-(\gamma) \text{ or } x > \gamma)(x + \sigma_-(\gamma))d\mu(x) = 0$$

$$(0 \cdot \infty = 0)$$

(d) $\sigma_-(\sigma_+(s)) < s$ for all s in $[0,\alpha]$.

PROOF. As usual, it suffices to consider σ_+.

(a) If $\gamma < \alpha$, then

$$\lim_{\lambda \to +\infty} \int (x - \lambda)I(x < -\gamma \text{ or } x > \lambda)d\mu(x) = -\infty.$$

Therefore $-\gamma_+(\lambda) < -\gamma$ for λ large enough, whence $\sigma_+(\gamma) < \infty$.

(b) is immediate from Lemma 3.2(b).

(c) If $\lambda < \sigma_+(\gamma)$, then $\gamma_+(\lambda) < \gamma$ and so (2.10) shows that

$$\int (x - \lambda)I(x < -\gamma \text{ or } x > \lambda)d\mu(x) > 0.$$

Let $\lambda \uparrow \sigma_+(\gamma) < \infty$, to get

$\int(x - \sigma_+(\gamma))I(x < -\gamma \text{ or } x > \sigma_+(\gamma))d\mu(x) > 0(-\infty \cdot 0 = 0).$

If $\sigma_+(\gamma) = \infty$, the above integrand is $(-\infty)I(x < -\gamma) < 0$ so that the integral must be zero. Assume therefore that $\sigma_+(\gamma) < \infty$ and let $\lambda > \sigma_+(\gamma)$. Then $\gamma_+(\lambda) > \gamma$ and so

$$\int(x - \lambda)I(x < -\gamma \text{ or } x > \lambda)d\mu(x) < 0 \text{ (by (2.10))}.$$

Let $\lambda \downarrow \sigma_+(\gamma)$ to get

$$\int(x - \sigma_+(\gamma))I(x < -\gamma \text{ or } x > \sigma_+(\gamma))d\mu(x) < 0.$$

This, together with the above converse inequality, proves (c).

(d) would follow from

$$\gamma_+(\gamma_-(s)) > s \quad \text{for } 0 < s < \alpha. \tag{3.10}$$

If $0 < s < \alpha$ and $\gamma_+(\gamma_-(s)) < s$, then lemma 3.2(d) with $a = s$ and $b = \gamma_-(s)$ shows that $s > \alpha$, a contradiction. This proves (3.10) for $0 < s < \alpha$. It is trivial for $s = 0$ and holds for $s = \alpha$ by left-continuity. □

NOTATION. $G(x) = \mu(-\infty,x]$, $K(\gamma) = \int_0^\gamma(1 - G(x))dx$, $H(\gamma) = \int_{-\gamma}^0 G(x)dx$,

$$f_\pm(t) = \exp\{\int_t^1(s + \sigma_\pm(s))^{-1}ds\} \quad (t > 0).$$

PROPOSITION 3.4. (a) $c_\pm = \lim_{\gamma\to 0+} \gamma f_\pm(\gamma)$ <u>exists and satisfies</u> $0 < c_\pm < 1$.

(b) (H,K) <u>satisfies the integral equations</u>

$$(3.11)\quad H(\gamma) = c_+G(0-)f_+(\gamma)^{-1}$$

$$+ \int_0^\gamma K(\sigma_+(s))df_+(s)f_+(\gamma)^{-1},\ \gamma > 0$$

$$(3.12)\quad K(\gamma) = c_-(1 - G(0))f_-(\gamma)^{-1}$$

$$+ \int_0^\gamma H(\sigma_-(s))df_-(s)f_-(\gamma)^{-1},\quad \gamma > 0.$$

PROOF. As H and f_+ are constant on $\{\gamma: \sigma_+(\gamma) = \infty\}$, it suffices to consider (3.11) for $0 < \gamma$ such that $\sigma_+(\gamma) < \infty$.

Integrate (3.8) by parts to see that for γ as above,

$$(-\gamma - \sigma_+(\gamma))G(-\gamma) - \int I(x < -\gamma)G(x)dx$$

$$- \int I(x > \sigma_+(\gamma))(G(x) - 1)dx = 0$$

$$\Rightarrow (\gamma + \sigma_+(\gamma))^{-1}\left(\int_0^\infty x\ dG(x) - K(\sigma_+(\gamma))\right)$$

$$= G(-\gamma) + (\gamma + \sigma_+(\gamma))^{-1}\left(\int_{-\infty}^0 -x\ dG(x) - H(\gamma)\right)$$

$$\Rightarrow\ -(\gamma + \sigma_+(\gamma))^{-1}K(\sigma_+(\gamma)) = G(-\gamma) - (\gamma + \sigma_+(\gamma))^{-1}H(\gamma)$$

$$(3.13)\qquad f_+'(\gamma)K(\sigma_+(\gamma)) = \frac{d^-}{d\gamma}(Hf_+)(\gamma).$$

Note that

$$Hf_+(\gamma) < H(\gamma)\exp\{\int_\gamma^1 s^{-1}ds\} \to G(0-)\ \text{as as}\ \gamma \downarrow 0.$$

As $Hf_+(\gamma)$ increases as $\gamma \downarrow 0$ (by (3.13)), it follows that $L_+ \equiv \lim_{\gamma\to 0+} Hf_+(\gamma)$ exists and belongs to $(0,G(0-)]$. We can now integrate (3.13) and conclude that

$$\int_0^\gamma K(\sigma_+(s))df_+(s) = Hf_+(\gamma) - L_+$$

for $0 < \gamma$ such that $\sigma_+(\gamma) < \infty$ and hence for all $\gamma > 0$. To obtain (3.11), simply note that

$$\lim_{\gamma\to 0+} \gamma f_+(\gamma) = \lim_{\gamma\to 0+} \gamma H(\gamma)^{-1} \lim_{\gamma\to 0+} H(\gamma)f_+(\gamma)$$

$$= G(0-)^{-1}L_+ \equiv c_+ \in (0,1].$$

The rest of (a) and (3.12) follow upon replacing μ with $\tilde{\mu}$.□

PROPOSITION 3.5. <u>If</u> μ_1, μ_2 <u>are mean-zero laws such that</u> $\gamma_\pm^{\mu_1} = \gamma_\pm^{\mu_2}$ <u>and</u> $\mu_1(\{0\}) = \mu_2(\{0\})$, <u>then</u> $\mu_1 = \mu_2$.

PROOF. Let $G_i(x) = \mu_i(-\infty,x]$, define H_i and K_i as above but with G_i in place of G, and write $\gamma_\pm$, $\sigma_\pm$, and $f_\pm$ for $\gamma_\pm^{\mu_i}$, $\sigma_\pm^{\mu_i}$, and $f_\pm^{\mu_i}$, respectively $(i = 1,2)$. Note that $\alpha^{\mu_i} = \gamma_+^{\mu_i}(\infty)$ and $\beta^{\mu_i} = \gamma_-^{\mu_i}(\infty)$, so we write α and β for α^{μ_i} and β^{μ_i}, respectively. (3.11) and (3.12) become

$$(3.14)_i \quad H_i(\gamma) = c_+G_i(0-)f_+(\gamma)^{-1}$$

$$+ \int_0^\gamma K_i(\sigma_+(s))df_+(s)f_+(\gamma)^{-1}, \quad \gamma > 0$$

$$(3.15)_i \quad K_i(\gamma) = c_-(1 - G_i(0))f_-(\gamma)^{-1}$$

$$+ \int_0^{\gamma} H_i(\sigma_-(s))df_-(s)f_-(\gamma)^{-1}, \gamma > 0.$$

Proposition 3.4(a), together with $f_-(u) > u^{-1} \wedge 1$, shows there is a $K > 0$ such that $f_-(u) > K^{-1}u^{-1}$ for all $u > 0$. Therefore if $\varepsilon < 1 < \gamma$, then we have

$$\int_{\varepsilon}^{\gamma} f_-(\sigma_+(s))^{-1}d(-f_+(s))$$

$$< K \int_{\varepsilon}^{\gamma} \sigma_+(s)f_+(s)(s + \sigma_+(s))^{-1}I(\sigma_+(s) < \infty)ds$$

$$< K \int_{\varepsilon}^{1} \sigma_+(s)(s^2 + s\sigma_+(s))^{-1}I(\sigma_+(s) < \infty)ds$$

$$+ K \int_{1}^{\gamma} \sigma_+(s)(s + \sigma_+(s))^{-1}I(\sigma_+(s) < \infty)ds$$

$$< -K \log(f_+(\varepsilon)\varepsilon) + K \int_{1}^{\gamma} I(\sigma_+(s) < \infty)ds$$

$$\to -K \log c_+ + K \int_{1}^{\gamma} I(\sigma_+(s) < \infty)ds \quad \text{as } \varepsilon \to 0.$$

Therefore we may define continuous, non-decreasing functions on $[0,\infty)$ by

$$g_+(\gamma) = \int_0^{\gamma} f_-(\sigma_+(u))^{-1}d(-f_+(u))$$

and symmetrically,

$$g_-(\gamma) = \int_0^{\gamma} f_+(\sigma_-(u))^{-1}d(-f_-(u)).$$

Substitute $(3.15)_i$ into $(3.14)_i$ to get

$$f_+(\gamma)H_i(\gamma) = c_+G_i(0-) + c_-(G_i(0) - 1)g_+(\gamma)$$

$$+ \int_0^\gamma \int_0^{\sigma_+(s)} H_i(\sigma_-(u))f_+(\sigma_-(u))\, dg_-(u)dg_+(s).$$

Take differences and recall that $\Delta G_1(0) = \Delta G_2(0)$ to see that

$$(3.16) \quad f_+(\gamma)(H_1(\gamma) - H_2(\gamma))$$

$$= (G_1(0) - G_2(0))(c_+ + c_-g_+(\gamma))$$

$$+ \int_0^\gamma \int_0^{\sigma_+(s)} (H_1(\sigma_-(u)) - H_2(\sigma_-(u)))f_+(\sigma_-(u))dg_-(u)dg_+(s).$$

Assume $G_1(0) > G_2(0)$. Then $\sigma_-(\sigma_+(s)) < s$ for $s < \alpha$ (Lemma 3.3(d)) and (3.16) show that $H_1(\gamma) > H_2(\gamma)$ for $\gamma < \alpha$ and hence for all $\gamma > 0$ because $H_i(\gamma) = H_i(\gamma \wedge \alpha)$. Let $\gamma \to +\infty$ in (3.16) to see that

$$(3.17) \quad \int_{-\infty}^0 -x\, dG_1(x) - \int_{-\infty}^0 -x\, dG_2(x) = H_1(\infty) - H_2(\infty) > 0.$$

Take differences in $(3.15)_i$ to get

$$(3.18) \quad K_2(\gamma) - K_1(\gamma) = (G_1(0) - G_2(0))c_-f_-(\gamma)^{-1}$$

$$+ \int_0^\gamma (H_1 - H_2)(\sigma_-(s))d(-f_-)(s)f_-(\gamma)^{-1}.$$

Letting $\gamma \to \infty$, we obtain

$$(3.19) \quad \int_0^\infty xdG_2(x) - \int_0^\infty xdG_1(x)$$

$$> (G_1(0) - G_2(0))c_-f_-(\gamma)^{-1} > 0.$$

Add (3.17) and (3.19) and conclude that $\int_{-\infty}^{\infty} x\, d(G_2 - G_1)(x) > 0$, contradicting the fact that G_1 and G_2 have mean zero. Hence our original assumption was false and we may conclude that $G_1(0) = G_2(0)$. (3.16) simplifies to

$$(3.20)\qquad f_+(\gamma)(H_1(\gamma) - H_2(\gamma)) = \int_0^{\gamma}\int_0^{\sigma_+(s)} (H_1(\sigma_-(u)) - H_2(\sigma_-(u)))f_+(\sigma_-(u))dg_-(u)dg_+(s).$$

Proposition 3.4(a) shows that

$$M(u) \equiv \sup_{0<\gamma<u} f_+(\gamma)|H_1(\gamma) - H_2(\gamma)| < \infty, \text{ for all } u > 0,$$

and (3.20) implies

$$M(u) < \int_0^u M(\sigma_-(\sigma_+(s)))g_-(\sigma_+(s))dg_+(s)$$

$$< g_-(\sigma_+(u)) \int_0^u M(s)dg_+(s) \quad \text{for } u < \alpha, \text{ by Lemma 3.3(d).}$$

An appropriate version of Gronwall's lemma shows that $M(u) = 0$ on $[0,\alpha) \subset \{\gamma : \sigma_+(\gamma) < \infty\}$ (Lemma 3.3(a)). As $H_i(\gamma) = H_i(\gamma \wedge \alpha)$, we have proved $H_1 = H_2$ and hence $K_1 = K_2$ by (3.18) and the fact that $G_1(0) = G_2(0)$. Differentiate to see that $G_1 = G_2$. □

LEMMA 3.6. $B(T_b) = \begin{cases} -\gamma_+(M(T_b)) = -m(T_b), & \underline{\text{if}}\ B(T_b) < 0 \\ \gamma_-(m(T_b)) = M(T_b), & \underline{\text{if}}\ B(T_b) > 0 \end{cases}$

PROOF. If $T_b = 0$, the result is obvious. By symmetry it suffices to consider the case when $B(T_b) < 0$ and $T_b > 0$. By definition there are $t_n \downarrow T_b$ such that $B(t_n) < -\gamma_+(M(t_n))$ and $M(t_n) = M(T_b)$ a.s. (the latter because $M(T_b) > 0 > B(T_b)$ a.s.). Therefore $B(t_n) < -\gamma_+(M(T_b))$ and we can let $n \to \infty$ to see that $B(T_b) < -\gamma_+(M(T_b))$ a.s. If $u_n \uparrow\uparrow T_b$, then for a.a. ω and large enough n we have

$$B(u_n) > -\gamma_+(M(u_n)) = -\gamma_+(M(T_b)).$$

Let $n \to \infty$ in the above to obtain $B(T_b) > -\gamma_+(M(T_b))$. This proves $-\gamma_+(M(T_b) = B(T_b)$. If $0 < t < T_b$, then

$$B(t) > -\gamma_+(M_t) > -\gamma_+(M(T_b)) = B(T_b),$$

and therefore $B(T_b) = -m(T_b)$. □

NOTATION. $\delta = \delta^\mu = \sup\{x > 0: \mu[0,x) = 0\}$

$$-\varepsilon = -\varepsilon^\mu = \inf\{x < 0: \mu(x,0] = 0\}.$$

THEOREM 3.7. T_b <u>is an embedding of</u> μ.

PROOF. **case 1.** $-\infty < -\alpha < -\varepsilon < 0 < \delta < \beta < \infty$.

In this case $-\alpha < -\gamma_+(\lambda) < -\varepsilon$ and $\delta < \gamma_-(\lambda) < \beta$ for $\lambda > 0$ and hence $B(t \wedge T_b)$ is uniformly bounded, $0 < T_b < \infty$, and $B(T_b) \neq 0$ a.s. Let ν denote the law of $B(T_b)$ and continue to write $\gamma_\pm$ for $\gamma_\pm^\mu$. We will use Proposition 3.5

to show $\nu = \mu$. The previous lemma shows

$$(3.21) \quad \{B(T_b) > \lambda \text{ or } B(T_b) < -\gamma_+(\lambda)\} \subset \{M(T_b) > \lambda\}$$

$$\subset \{B(T_b) > \lambda \text{ or } B(T_b) < -\gamma_+(\lambda)\} \text{ for all } \lambda > 0,$$

which in turn implies

$$\int I(B(T_b) > \text{ or } B(T_b) < -\gamma_+(\lambda))B(T_b)dP$$

$$> \int I(M(T_b) > \lambda)B(T_b)dP$$

$$= \lambda\ P(M(T_b) > \lambda) \text{ (optional stopping)}$$

$$> \lambda\ P(B(T_b) > \lambda \text{ or } B(T_b) < -\gamma_+(\lambda)).$$

It follows immediately that $\gamma_+^\nu(\lambda) < \gamma_+(\lambda)$ for all $\lambda > 0$. If $\lambda < \beta^\nu$ and $\lambda' \in (\lambda, \beta^\nu)$, then

$$\int I(B(T_b) > \lambda' \text{ or } B(T_b) < -\gamma_+(\lambda))B(T_b)dP$$

$$< \int I(M(T_b) > \lambda)B(T_b)dP$$

$$- \int I(\lambda < B(T_b) < \lambda')B(T_b)dP \text{ (by (3.21))}$$

$$< \lambda\ P(M(T_b) > \lambda) - \lambda\ P(\lambda < B(T_b) < \lambda')$$

$$< \lambda'\ P(B(T_b) > \lambda' \text{ or } B(T_b) < -\gamma_+(\lambda)),$$

the last by (3.21) and the fact that $\lambda' < \beta^{\nu}$. This shows that for λ, λ' as above, $\gamma_{+}^{\nu}(\lambda') > \gamma_{+}(\lambda)$. First let $\lambda' \downarrow\downarrow \lambda$ and then take limits from below (using the left continuity of γ_{+}^{ν}, γ_{+}) to see that $\gamma_{+}^{\nu}(\lambda) > \gamma_{+}(\lambda)$ for $\lambda < \beta^{\nu}$. We have therefore shown

$$\gamma_{+}^{\nu}(\lambda) = \gamma_{+}(\lambda) \qquad \text{for } 0 < \lambda < \beta^{\nu} < \beta$$

(the last inequality is clear because $\gamma_{-} < \beta$), and symmetrically,

$$\gamma_{-}^{\nu}(\lambda) = \gamma_{-}(\lambda) \qquad \text{for } 0 < \lambda < \alpha^{\nu} < \alpha.$$

In particular, $\gamma_{+}(\beta^{\nu}) = \gamma_{+}^{\nu}(\beta^{\nu}) = \alpha^{\nu}$ and $\gamma_{-}(\alpha^{\nu}) = \gamma_{-}^{\nu}(\alpha^{\nu}) = \beta^{\nu}$, results that allow us to apply Lemma 3.2(d) and conclude that $\mu([-\alpha^{\nu},\beta^{\nu}]^{c}) = 0$. This means $\alpha^{\nu} = \alpha$, $\beta^{\nu} = \beta$ and hence $\gamma_{\pm}^{\nu}(\lambda) = \gamma_{\pm}(\lambda)$ for all $\lambda > 0$. As $\mu(\{0\}) = \nu(\{0\}) = 0$, Proposition 3.5 implies $\nu = \mu$.

case 2. $-\infty < -\alpha$, $\beta < \infty$, $\mu(\{0\}) = 0$.

Choose $\varepsilon_n \downarrow\downarrow 0$ and let $K_n = \mu[-\varepsilon_n,\varepsilon_n]$,

$$m_n = \int I(-\varepsilon_n < x < \varepsilon_n)x\, d\mu(x)/K_n \quad (0/0 = 0).$$

Pick r_n in [0,1] such that $m_n = r_n(-\varepsilon_n) + (1 - r_n)\varepsilon_n$ and let

$$\mu_n(A) = \mu([-\varepsilon_n,\varepsilon_n]^{c} \cap A) + K_n r_n \delta_{-\varepsilon_n}(A) + K_n(1 - r_n)\delta_{\varepsilon_n}(A).$$

μ_n is a mean-zero probability satisfying the conditions of case 1. Therefore, if we write T_n for $T_b^{\mu_n}$ and $\gamma_\pm^n$ for $\gamma_\pm^{\mu_n}$, then $\mathscr{L}(B(T_n)) = \mu_n$. If $\lambda > \varepsilon_n$ and $\gamma_+(\lambda) > \varepsilon_n$ then

$$\int I(x < -\varepsilon_n \text{ or } x > \lambda)(x - \lambda)d\mu_n(x) < 0$$

and so

$$-\gamma_+^n(\lambda) = \sup\{y < -\varepsilon_n : \int I(x < y \text{ or } x > \lambda)(x - \lambda)d\mu_n(x) > 0\}$$

$$= \sup\{y < -\varepsilon_n : \int I(x < y \text{ or } x > \lambda)(x - \lambda)d\mu(x) > 0\}$$

$$= -\gamma_+(\lambda) \qquad\qquad (\gamma_+(\lambda) > \varepsilon_n).$$

By symmetry we have

(3.22) $\quad \gamma_\pm^n(\lambda) = \gamma_\pm(\lambda)$ if $\lambda > \varepsilon_n$ and $\gamma_\pm(\lambda) > \varepsilon_n$.

Choose $q_n \downarrow\downarrow 0$ such that $\mu[-q_n, q_n] < 2^{-n}$ and then $p_n \downarrow\downarrow 0$ such that

(3.23) $\quad P(\max(T_B(p_n), T_B(-p_n)) > T_{|B|}(q_n)) < 2^{-n}$.

As $\gamma_\pm(\lambda) > 0$ if $\lambda > 0$, (3.22) shows that we may choose $\{\varepsilon_n\}$ so that

$$\gamma_\pm^n(\lambda) = \gamma_\pm(\lambda) \text{ for } \lambda > p_n \text{ and } \mu = \mu_n \text{ on } [-q_n, q_n]^c.$$

This shows that

$$T_n' \equiv \inf\{t > \max(T_B(p_n),T_B(-p_n)):$$

$$B_t \notin (-\gamma_+^n(M_t),\gamma_-^n(m_t))\}$$

(3.24) $$= \inf\{t > \max(T_B(p_n),T_B(-p_n)):$$

$$B_t \notin (-\gamma_+(M_t),\gamma_-(m_t))\},$$

and therefore

$$P(T_n \neq T_n') < P(T_n < \max(T_B(p_n),T_B(-p_n)))$$

$$< P(T_n < T_{|B|}(q_n)) + 2^{-n} \qquad \text{(by (3.23))}$$

$$< \mu_n[-q_n,q_n] + 2^{-n}$$

$$= \mu[-q_n,q_n] + 2^{-n} < 2^{-n+1}.$$

The Borel-Cantelli lemma implies

(3.25) $T_n = T_n'$ for large enough n a.s.

(3.24) shows that $T_n' \downarrow T_\infty' > T_b$ a.s. Let $t \in (0,T_\infty')$ and choose n large enough so that $\max(T_B(p_n), T_B(-p_n)) < t$. We must have $B_t \in (-\gamma_+(M_t),\gamma_-(m_t))$ because $t < T_n'$. This shows that $T_\infty' < T_b$ and hence (3.25) shows that $T_n \to T_\infty' = T_b$ a.s. Therefore $B(T_n) \to B(T_b)$ a.s. This shows that $\mathscr{L}(B(T_b)) = \mu$ because $\mathscr{L}(B(T_n)) = \mu_n \overset{w}{\to} \mu$. T_b is an embedding of μ because $B(T_b \wedge t)$ is bounded.

case 3. $\mu(\{0\}) = 0$.

Let $-\alpha_n \downarrow\downarrow -\alpha$ $(\alpha_n > 0)$ and define

$$\beta_n = \inf\{\lambda > 0: \int_{-\infty}^{0} (-x) \wedge \alpha_n d\mu(x) = \int_{0}^{\infty} x \wedge \lambda \, d\mu(x)\}.$$

Then $0 < \beta_n \uparrow \beta$, $\beta_n < \beta$ and

$$\mu_n(A) \equiv \mu(-\infty,-\alpha_n]\delta_{\alpha_n}(A) + \mu(A \cap (-\alpha_n,\beta_n)) + \mu[\beta_n,\infty)\delta_{\beta_n}(A).$$

is a mean-zero probability with compact support. Therefore if we write T_n for $T_b^{\mu_n}$ and $\gamma_\pm^n$ for $\gamma_\pm^{\mu_n}$, then $\mathscr{L}(B(T_n)) = \mu_n$. An argument similar to that given in case 2 shows that

$$\gamma_+^n(\lambda) = \gamma_+(\lambda) \text{ if } \lambda < \beta_n \text{ and } \gamma_+(\lambda) < \alpha_n \tag{3.26}$$

$$\gamma_-^n(\lambda) = \gamma_-(\lambda) \text{ if } \lambda < \alpha_n \text{ and } \gamma_-(\lambda) < \beta_n. \tag{3.27}$$

Note that since μ and μ_n are mean-zero laws that agree on $(-\alpha_n,\beta_n)$, one has

$$\int I(x < -\alpha_n \text{ or } x > \beta_n)x \, d\mu(x)$$

$$= \int I(x < -\alpha_n \text{ or } x > \beta_n)x \, d\mu_n(x)$$

$$= -\alpha_n\mu(-\infty,-\alpha_n] + \beta_n\mu[\beta_n,\infty) < \beta_n(\mu(-\infty,-\alpha_n] + \mu[\beta_n,\infty)).$$

This shows that $\gamma_+(\beta_n) > \alpha_n$ and hence $\gamma_+(\lambda) > \alpha_n > \gamma_+^n(\lambda)$ for $\lambda > \beta_n$ or $\gamma_+(\lambda) > \alpha_n$. Combine this with (3.26) to conclude that $\gamma_+(\lambda) > \gamma_+^n(\lambda)$ for all $\lambda > 0$. As μ may be replaced by μ_{n+1} in the definition of μ_n, this in fact shows $\gamma_\pm^n < \gamma_\pm^{n+1} < \gamma_\pm$ and therefore $T_n \uparrow T_\infty < T_b$, and $\gamma_\pm^n \uparrow \gamma_\pm^\infty < \gamma_\pm$. Fix $\lambda > 0$ and choose $\gamma > \gamma_+^\infty(\lambda)$ such that $\mu(\{-\gamma\}) = \mu_n(\{-\gamma\}) = 0$ for each n. $\mu_n \overset{w}{\to} \mu$ and

$$(3.28) \qquad \int |x| d\mu_n \to \int |x| d\mu \qquad \text{as } n \to \infty.$$

It follows that

$$\int I(x < -\gamma \text{ or } x > \lambda)(x - \lambda) d\mu(x)$$

$$= \lim_{n\to\infty} \int I(x < -\gamma \text{ or } x > \lambda)(x - \lambda) d\mu_n(x)$$

$$> 0 \ (\gamma > \gamma_+^\infty(\lambda) > \gamma_+^n(\lambda)).$$

Therefore $\gamma > \gamma_+(\lambda)$ and letting $\gamma \downarrow \gamma_+^\infty(\lambda)$ we see that $\gamma_+^\infty(\lambda) > \gamma_+(\lambda)$. By symmetry we have shown that $\lim_{n\to\infty} \gamma_\pm^n(\lambda) = \gamma_\pm(\lambda)$ for all $\lambda > 0$.

We now show $T_b < \infty$ a.s. If α or β is finite this is obvious because $T_b < \min(T_B(-\alpha), T_B(\beta))$. Assume therefore $\alpha = \beta = \infty$. (3.26) and (3.27) show $T_b = T_n$ if $M(T_n) < \beta_n \wedge \sigma_+(\alpha_n) \equiv a_n$ and $m(T_n) < \alpha_n \wedge \sigma_-(\beta_n) \equiv b_n$. These latter conditions are implied by

$$-b_n \vee (-\gamma_+(a_n)) < B(T_n) < a_n \wedge \gamma_-(b_n)$$

(Lemma 3.6). Therefore

$$P(T_b = T_n) > P((-b_n \vee (-\gamma_+(a_n)) < B(T_n) < a_n \wedge \gamma_-(b_n))$$

$$= \mu((-\gamma_+(a_n)) \vee (-b_n), \gamma_-(b_n) \wedge a_n)$$

$$\to 1 \quad \text{as } n \to \infty$$

because $\sigma_\pm(\infty) = \gamma_\pm(\infty) = \infty$ if $\alpha = \beta = \infty$. The fact that $T_n < \infty$ for all n a.s. now shows that $T_b < \infty$ a.s., and hence $T_\infty < \infty$ a.s. also.

If $B(T_n) = \gamma_-^n(m_{T_n})$ for infinitely many n then, taking limits, we see that $B(T_\infty) > 0$ a.s. (recall $P(T_n = 0) = 0$ by case 2). Therefore if $B(T_\infty) < 0$, Lemma 3.6 shows

$$B(T_\infty) = \lim_{n\to\infty} - \gamma_+^n(M(T_n)) = -\gamma_+(M(T_\infty)).$$

The last equality holds because $\gamma_+^n \uparrow \gamma_+$ and the limit is left-continuous (Lemma 3.2). This result, together with a similar conclusion if $B(T_\infty) > 0$, shows that $T_\infty > T_b$. Therefore $T_n \uparrow T_b$ and so $\mathscr{L}(B(T_b))$ is the weak limit of $\mathscr{L}(B(T_n)) = \mu_n$, namely μ. (3.28) implies that $\{B(T_n): n \in \mathbf{N}\}$, and hence $\{B(T_b \wedge t): t > 0\}$, is uniformly integrable.

case 4. General μ.

Assuming without loss of generality that $\mu(\{0\}) < 1$, let $\nu(A) = \mu(A|\mathbf{R} - \{0\})$. Then $\gamma_\pm^\nu = \gamma_\pm^\mu$ and therefore $T_b^\nu = T_d^\mu$ is an embedding of ν by the previous case. This implies

$$\mathscr{L}(B(T_b)) = \mu(\{0\})\delta_0 + (1 - \mu(\{0\}))\, \mathscr{L}(B(T_d^\mu))$$

$$= \mu(\{0\})\delta_0 + (1 - \mu(\{0\}))\nu = \mu.$$

The fact that $T_b^\mu \leq T_b^\nu$ shows that $\{B(t \wedge T_b^\mu): t \geq 0\}$ is uniformly integrable. □

THEOREM 3.8. <u>Let</u> T <u>be any embedding of</u> μ.

(a) <u>For all</u> $\lambda > 0$,

(i) $P(M(T_b) \geq \lambda) = \mu_+(\lambda) \leq P(M(T) \geq \lambda)$

(ii) $P(m(T_b) \geq \lambda) = \mu_-(\lambda) \leq P(m(T) \geq \lambda)$

(iii) $P(B^*(T_b) \geq \lambda) = \max\{\mu_+(\lambda), \mu_-(\lambda),$
$\mu(|x| \geq \lambda)\} \leq P(B^*(T) \geq \lambda).$

(b) <u>If</u> $E(M(T) + m(T)) = E(M(T_b) + m(T_b))$, <u>then</u> $T = T_d^\mu$ <u>on</u> $\{T > 0\}$ a.s. <u>and</u> $P(T = 0) = \mu(\{0\})$. <u>In particular, if</u> $\mu(\{0\}) = 0$, <u>then</u> $T = T_b^\mu$ a.s.

PROOF. (a) (3.21) shows that we may use Theorem 2.5(b) to conclude that $P(M(T_b) = \lambda) \geq \mu_+(\lambda)$. By symmetry one gets $P(m(T_b) > \lambda) = \mu_-(\lambda)$. The inequalities in (i) and (ii) are immediate from Theorem 2.5(b).

Lemma 3.6 implies that for $\lambda > 0$,

$$P(M(T_b) \geq \lambda,\ m(T_b) < \lambda)$$

$$= P(M(T_b) \geq \lambda,\ m(T_b) < \lambda,\ B(T_b) > 0)$$

$$+ P(M(T_b) \geq \lambda,\ m(T_b) < \lambda,\ B(T_b) < 0)$$

$$\leq P(\lambda \leq B(T_b) \leq \gamma_-(\lambda)) + P(-\lambda \leq B(T_b) \leq -\gamma_+(\lambda)). \tag{3.29}$$

Replace B with -B and μ with $\tilde{\mu}$ to see that

$$(3.30) \quad P(m(T_b) > \lambda,\ M(T_b) < \lambda) \leq P(-\gamma_+(\lambda) \leq B(T_b) < -\lambda)$$

$$+ P(\gamma_-(\lambda) \leq B(T_b) < \lambda).$$

To prove (iii) we consider four cases.

case 1. $\gamma_-(\lambda) < \lambda \leq \gamma_+(\lambda)$.

$$P(B^*(T_b) > \lambda) = P(m(T_b) > \lambda) + P(M(T_b) > \lambda,\ m(T_b) < \lambda)$$

$$= \mu_-(\lambda),$$

by (3.29) and (ii).

case 2. $\gamma_+(\lambda) < \lambda \leq \gamma_-(\lambda)$.

Use (3.30) as above to see that $P(B^*(T_b) > \lambda) = \mu_+(\lambda)$.

case 3. $\lambda \leq \gamma_+(\lambda)$ and $\lambda \leq \gamma_-(\lambda)$.

Lemma 3.6 shows that

$$P(B^*(T_b) > \lambda) \leq P(B(T_b) > \min(\lambda,\gamma_-(\lambda))$$

$$+ P(B(T_b) < \max(-\lambda,-\gamma_+(\lambda))) = P(|B(T_b)| > \lambda)$$

$$= \mu(|x| > \lambda).$$

case 4. $\lambda > \gamma_+(\lambda)$ and $\lambda > \gamma_-(\lambda)$.

Choose $\lambda' < \lambda$ such that $\lambda' > \gamma_\pm(\lambda')$. Lemma 3.2(d), with $a = b = \lambda'$, shows that $\mu([-\lambda',\lambda']^c) = 0$ and therefore $\gamma_\pm(t) < \lambda'$ for all $t > 0$. This in turn implies $P(B^*(T_b) > \lambda) < P(B^*(T_b) > \lambda') = 0$.

(iii) follows easily from the above, and (i) and (ii).

(b) If $E(M(T) + m(T)) = E(M(T_b) + m(T_b))$, then (a) shows that

$$P(M(T) > \lambda) = P(M(T_b) > \lambda) = \mu_+(\lambda)$$

$$P(m(T) > \lambda) = P(m(T_b) > \lambda) = \mu_-(\lambda)$$

for all $\lambda > 0$. Theorem 2.5(b) gives

$$\{B_T > \lambda \text{ or } B_T < -\gamma_+(\lambda)\} \subset \{M_T > \lambda\}$$

$$\subset \{B_T > \lambda \text{ or } B_T < -\gamma_+(\lambda)\} \text{ for all rational } \lambda > 0 \text{ a.s.}$$

Approximating M_T from below by rationals in the latter inclusion, we obtain

(3.31) $\quad B_T = M_T$ or $B_T < -\gamma_+(M_T)$ a.s.

Symmetrically we have

(3.32) $\quad B_T = -m_T$ or $B_T > \gamma_-(m_T)$ a.s.

(3.31) and (3.32), together with Lemma 3.2(c), show that

(3.33) if $T > 0$, then $T \geq T_d$ a.s.

(3.34) $\{T = 0\} = \{B_T = 0\}$ a.s.,

whence $P(T = 0) = \mu(\{0\})$. Assuming, without loss of generality that $\mu(\{0\}) < 1$, let $Q(A) = P(A|T > 0) = P(A|B_T \neq 0)$ and $\nu(C) = \mu(C|\{0\}^c)$. Then B_t is a Q-Brownian motion and T (on$(\Omega, \mathscr{F}, Q)$) is an embedding of ν. (B, T_d) is independent of $\{T = 0\}$ because T_d is measurable function of B. Therefore

$$Q(B(T_d) \in A) = P(B(T_d) \in A) = \nu(A)$$

(recall from case 4 of Theorem 2.7 that $T_d^{\mu} = T_b^{\nu}$). Hence T and T_d are both embeddings of ν (on$(\Omega, \mathscr{F}, Q)$) and $T \geq T_d$ Q-a.s. by (3.33). This implies $T = T_d$ Q-a.s. (see Chacon-Ghoussoub (1979), p. 27)) and therefore $T = T_d$ a.s. on $\{T > 0\}$. If $\mu(\{0\}) = 0$, then $P(T = 0) = P(T_b = 0) = 0$ and one has $T = T_b$ a.s. □

REMARKS. 1. (b) shows that T_b is the essentially unique embedding that minimizes $E(M_T + m_T)$ over all embeddings. The corresponding uniqueness theorem for $E(B_T^*)$ is false. Indeed, if μ is symmetric, then the Skorokhod time T_s, the filling scheme T_c, and T_b all satisfy $\mathscr{L}(B_T^*) = \mathscr{L}(|B_T|)$.

2. It is now of some interest to compute T_b in some specific cases. If μ assigns probability 1/4 to each of

the points ±2, ±1 then $\gamma_{\pm}^{\mu}(\lambda) = 1 + I(\lambda > 1/3)$ (for $\lambda > 0$). If $\tau(A) = \inf\{t > 0: B_t \in A\}$, then

$$T_b = \begin{cases} \tau(1) & \text{if } m(\tau(1)) < 1/3 \\ \tau(-1) & \text{if } M(\tau(-1)) < 1/3 \\ \tau(\{\pm 2\}) & \text{otherwise.} \end{cases}$$

If μ is the uniform distribution on [-1,1], then

$$\gamma_{\pm}^{\mu}(\lambda) = \begin{cases} 2\sqrt{\lambda} - \lambda, & \lambda < 1 \\ 1 \quad , & \lambda > 1 \end{cases},$$

and so

$$T_b = \inf\{t > 0: B_t > 2\sqrt{m_t} - m_t \text{ or } B_t < -2\sqrt{M_t} + M_t\}.$$

These results are most impressive if you start with the definition of T_b and ask an unsuspecting friend for the law of $B(T_b)$.

3. The existence of a unique, and fairly explicit, extremal embedding should be compared to section 4 of Davis (1980), where a similar question is considered for rearrangements of an integrable function f on the unit circle. Here the problem is to find a rearrangement of f of minimal H^p-norm for $0 < p < 2$ and maximal H^p-norm for $2 < p < \infty$. In this setting the extremal problem is harder to solve because one must work with a restricted class of continuous martingales. Indeed there need not be an extremal rearrangement (in the above sense) in general, and

even if one exists, it may be rather difficult to describe explicitly. Note also the extremality properties of T_b are stronger than those of the extremal rearrangement obtained by Davis. This is essentially caused by the restriction $\int \tilde{f}^2 dm = \int f^2 dm$ where $\tilde{f}$ is the conjugate function of f.

4. If ν is a second mean-zero probability on **R**, write $\nu \vdash \mu$ if there is a Brownian motion B and a stopping time T such that $\mathscr{L}(B_0) = \nu$, $\mathscr{L}(B_T) = \mu$ and $B_{t \wedge T}$ is uniformly integrable. Such a T is an embedding from ν to μ, and is called an H^1-embedding from ν to μ if, in addition, $E(B_T^*) < \infty$. It is easy to see that

(3.35) $\quad \nu \vdash \mu \iff \mu = \nu * \eta$ for some probability η.

If $h_\nu(t) = \int e^{itx} d\nu(x)$ and

(3.36) $\quad \{t : h_\nu(t) \neq 0\}$ is dense in **R**,

then the law, η, appearing in (3.35) is unique because $h_\eta(t) = h_\mu(t)/h_\nu(t)$ on a dense set of t.

Let $\tilde{B}$ be an $\{\mathscr{F}_t\}$-Brownian motion starting at zero, B_0 an $\mathscr{F}_0$-measurable r.v. with law ν (a mean-zero law), and $B_t = B_0 + \tilde{B}_t$. Assume (3.35) and let $T_b = \tilde{T}_b^\eta$ denote the embedding of η in $\tilde{B}$ considered in Theorem 3.7 (η is some fixed law obtained from (3.35)). We may, and shall, assume $(\tilde{B}, T_b)$ is independent of B_0. Then

$$\mathscr{L}(B(T_b)) = \mathscr{L}(B_0 + \tilde{B}(\tilde{T}_b^\eta)) = \nu * \eta = \mu,$$

and hence T_b is an embedding from ν to μ.

Let T be any embedding from ν to μ and let $\mathscr{L}(B_T - B_0) = \eta$. If $T_b = \tilde{T}_b^\eta$, as above, then

$$P(B_0 < \lambda < M_T) = \int I(x < \lambda)P(\tilde{M}_T > \lambda - x)d\nu(x)$$

$$(\tilde{M}_T = \inf_{s<T}\tilde{B}_s)$$

$$> \int I(x < \lambda)P(\tilde{M}_{T_b} > \lambda - x)d\nu(x) \text{ (Theorem 3.8)}$$

$$= P(B_0 < \lambda < M_{T_b}),$$

and therefore

$$(3.37) \quad P(M_T > \lambda) > P(M_{T_b} > \lambda) \text{ for all } \lambda > 0.$$

Similarly we have

$$(3.38) \quad P(m_T > \lambda) > P(m_{T_b} > \lambda) \text{ for all } \lambda > 0.$$

Assume (3.36). Then (3.37) and (3.38) holds for any embedding from ν to μ. If $E(M_T + m_T) = E(M_{T_b} + m_{T_b})$ then $E(\tilde{M}_T + \tilde{m}_T) = E(\tilde{M}_{T_b} + \tilde{m}_{T_b})$. The uniqueness of η in (3.35) shows that T is an embedding of η in $\tilde{B}$. Therefore Theorem 3.8(b) shows that $T = \tilde{T}_d^\eta$ on $\{T > 0\}$ a.s. (the ~ indicates the underlying Brownian motion is $\tilde{B}$) and in particular $T = T_b$ a.s. if $\eta(\{0\}) = 0$.

Finally the above remarks (especially (3.37), (3.38)) together with Theorems 3.8 and Lemma 2.6 prove

THEOREM 3.9. Let ν, μ be a mean-zero probabilities on **R**. There is an H^1-embedding from ν to μ iff there is a probability η such that $\mu = \nu * \eta$ and $H(\eta) < \infty$.

ACKNOWLEDGEMENT. I wish to thank Burgess Davis and John Walsh for explaining their results to me and also Paul-André Meyer and Michel Emery for an enjoyable visit to the University of Strasbourg, during which most of this work was done.

References

1. J. Azéma, M. Yor (1978a). Une solution simple au problème de Skorokhod. Séminaire de Probabilités XIII, p. 90-115. Lect. Notes in Math. 721, Springer-Verlag, Berlin, Heidelberg, New York.

2. J. Azéma, M. Yor (1978b). Le problème de Skorokhod: compléments à l'exposé précédent. Séminaire de Probabilités XIII, p. 626-633. Lect. Notes in Math. 721, Springer-Verlag, Berlin, Heidelberg, New York.

3. J. R. Baxter. Balayage in least time, unpublished manuscript.

4. D. Blackwell, L. E. Dubins (1963). A converse to the dominated convergence theorem, Illinois J. Math. 7, 508-514.

5. D. L. Burkholder, R. F. Gundy and M. L. Silverstein (1971). A maximal function characterization of the class H^p, Trans Amer. Math. Soc. 157, 137-153.

6. O. D. Cereteli (1976). A metric characterization of the set of functions whose conjugate functions are integrable, Bull. of the Academy of Sciences of the Georgian S.S.R. 81, 281-283 (in Russian).

7. P. Chacon(1985). Ph.D. thesis, U. of Washington.

8. R. Chacon, N. Ghoussoub (1979). Embeddings in Brownian motion, Annales de l'Institut Henri Poincaré, Section B 15, 287-292.

9. B. Davis (1980). Hardy spaces and rearrangements, Trans. Amer. Math. Soc. 261, 211-233.

10. B. Davis (1982). On the integrability of the ergodic maximal function, Studia Mathematica 73, 153-167.

11. L. E. Dubins, D. Gilat (1978). On the distribution of maxima of martingales, Proc. of Amer. Math. Soc. 68, 337-338.

12. H. Rost (1971). The stopping distributions of a Markov process, Inventiones Math. 14, 1-16.

13. A. Skorokhod (1965). Studies in the theory of random processes, Addison-Wesley, Reading.

14. P. Vallois (1982). Le probléme de Skorokhod sur R, une approche avec le temps local. Séminaire de Probabilités XVII, p. 227-239. Lect. Notes in Math. 986, Springer-Verlag, Berlin, Heidelberg, New York, Tokyo.

Edwin Perkins
Department of Mathematics
University of British Columbia
Vancouver, B.C.
V6T 1Y4
Canada

Seminar on Stochastic Processes, 1985
Birkhäuser, Boston, 1986

THINNESS AND HYPERTHINNESS

by

Z. R. Pop-Stojanovic

Introduction

Concepts of "small" or in some sense exceptional sets play essential roles in many parts of analysis, probability theory as well as in both classical and probabilistic potential theory. There are many examples: sets of first category, sets of measure zero, polar sets, are all examples of such "small" sets. In many situations one has to compare these sets: although a set of first category is small and so is a set of measure zero, these sets are not the same. In potential theory a concept of "smallness" of a set at a point is of special interest. More precisely, the concept of a set thin at a point is of major interest in potential theory. Originally, this notion arose in classical potential theory in conjuction with Dirichlet problem. Here, the ultimate characterization of regularity is a necessary and sufficient condition due to N. Wiener [7]. THe so-called Wiener's test gives necessary and sufficient conditions in terms of capacity for a point of a set to be irregular, i.e., for a set to be thin at a point. (See more in [3], [4], [5], [6]).

Thinness of a set at a point has its probabilistic counterpart as seen in Blumenthal-Getoor [1]. Here, both concepts agree. In classical case, M. Brelot [2] has introduced a concept of hyperthinness. Again, in the classical case both these concepts agree.

The purpose of this note is to show that under certain conditions in probabilistic potential theory thinness will imply hyperthinness.

Preliminaries

All notations in this paper are generally that of Blumenthal-Getoor [1]. Thus, $X = (\Omega, \mathcal{F}, \mathcal{F}_t, X_t, \theta_t, P^x)$ denotes a transient Hunt process on a locally compact second countable state space $(E, \mathcal{E})$. The family (P_t) denotes the transition semigroup of the process. Furthermore, we assume that:

1) (P_t) is strong Feller, i.e., $P_t f$ is continuous if f is bounded measurable function.

2) There is an excessive reference measure denoted by dx.

3) There is a potential density u:

$$Uf(x) = \int u(x,y)\, f(y)\, dy$$

such that the Riesz decomposition is valid, i.e., every excessive function s can be written as

$$s = h + \int u(x,y)\, \mu\, (dy)$$

where μ is an unique Radon measure, and h is "harmonic."

4) Hypothesis (B) holds.

5) Points are polar sets L.

The setting.

Relative to the process X we shall now introduct the following

DEFINITION 1. (Blumenthal-Getoor [1]). A set $A \subset E$ is thin at x if there exists a set $D \in \mathcal{E}^n$ such that $D \supset A$ and x is irregular for D, that is, $P^x(T = 0) = 0$.

REMARK 1. It can be shown (see the example 3.14 in [1], page 83) that for a given process a set may be thin at every point without being a thin set.

REMARK 2. Following the example 4.15 in [1], page 87, one can give the following equivalent characterization of the thinness of a set at a point: let K be a compact set, $K \in \mathcal{E}^n$, $x_o \in K$. Then, K is thin at x_o if there exists an excessive function t relative to the process X such that

$$\liminf_{\substack{x \to x_o \\ x \in K - \{x_o\}}} t(x) > t(x_o). \tag{1}$$

DEFINITION 2. (M. Brelot [2]). A compact set K is said to be <u>hyperthin at</u> x_o if there exists an excessive function t relative to the process X such that $t(x_o)$ is finite and $t(x) \to +\infty$, as $x \to x_o$, $x \in K - \{x_o\}$, or, equivalently,

$$(1') \qquad \liminf_{\substack{x \to x_o \\ x \in K-\{x_o\}}} t(x) = +\infty.$$

Note, that (1') implies that if a set K is hyperthin at x_o then it is also thin at x_o.

Therefore, a natural question arises: When does the thinness of a set at a point imply the hyperthinness of the same set at that point? In this paper we shall give conditions under which this implication holds.

Assume that X is a transient Hunt process introduced earlier and satisfying conditions 1) - 5). First, we establish the following

PROPOSITION <u>Let</u> K <u>be a compact set</u>, $K \subset E$, <u>and assume that</u> K <u>is thin at</u> x_o <u>relative to the process</u> X. <u>Then there exists a measure</u> μ <u>on</u> K <u>such that</u> $s = U\mu$ <u>is a bounded function and</u>

$$(2) \qquad s(x_o) < \liminf_{\substack{y \to x_o \\ y \in K-\{x_o\}}} s(y).$$

PROOF. Since K is thin at x_o relative to X, there exists an excessive function t such that (1) holds. We may assume that t is a bounded function. Then, from the assumption 3), it follows that

$$t = h + U\mu$$

for some Radon measure μ with h harmonic. Now, by using the assumption 1), i.e., the assumption that X is a strong Feller, it follows that a bounded harmonic function h is continuous. Thus, we may assume that $t = U\mu$, and take $s = t$. Q.E.D.

Having in mind the previous Proposition, suppose that $s = Um$, where s is bounded and a measure m is with compact support K. Let $x_o \in K$. Then, s is a class (D) potential, $s = U_A$, where by Meyer's energy formula one gets that

$$E^{\cdot}[A_{\infty}^{2}]$$

is a bounded excessive function. Here, A_{∞} denotes the additive functional of s.

Assume,

$$(*) \qquad \liminf_{\substack{y \to x_o \\ y \in K - \{x_o\}}} s(y) > s(x_o) + \xi, \ \xi > 0.$$

Put $p = E^{\cdot}[A_{\infty}^{2}]$. Then, $p = Ua$ for some measure a whose support is K. Indeed,

$$p = E^{\cdot}[\int_{0}^{\infty} [s(X_t) + s(X_t)_{-}]dA_t],$$

so one can see that the Revuz measure of p is absolutely continuous with respect to that of s. The measure a is a finite measure since a is with support on K and p is bounded, hence locally integrable. In particular,

$$\int s \, da < +\infty.$$

Now, given $\varepsilon > 0$, there exists a compact set $L \subset K$ such that $s|_L$ is continuous and

$$(m + a)(K - L) < \varepsilon.$$

Now L does not contain x_o because $s|_L$ is continuous and (*) holds at x_o. Since s is bounded, $U(m|_L)$ is also bounded and from 1), i.e., the strong Feller assumption, $U(m|_L)$ will be continuous at x_o. Finally, from 4), i.e., the hypothesis (B), one gets:

$$\liminf_{\substack{y \to x_o \\ y \in K - \{x_o\}}} U(m|_{K-L})(y) > U(m|_{K-L})L(x_o) + \xi.$$

Here, we are using the fact that points are not charged by Revuz measure of class (D) potentials. This follows from the assumption 5) that points are polar sets.

Now we can choose compact sets (L_i), $L_i \subset K$, such that:

$$\sum_i m(K - L_i) < +\infty,$$

$$\sum_i [U(a(K - L_i))(x_o)]^{1/2} < +\infty.$$

This can be done since:

$$\sum_i [m(K - L_i) + (a(K - L_i))^{1/2}] < +\infty.$$

Write:

$$K_i = K - L_i,\ \eta_i = m|_{K_i},\ s_i = U\mu_i.\ \text{Then, } s_i = U_{A^i},\ \text{say.}$$

Then:

$$E^{x_o}[(A^i_\infty)^2] = E^{x_o}[\int_0^\infty [s_i(X_t) + s_i(X_t)_-]dA^i_t]$$

$$= E^{x_o}[\int_0^\infty [s_i(X_t) + s_i(X_t)_-]1_{K_i}(X_t)dA^i_t]$$

$$\leq E^{x_o}[\int_0^\infty s(X_t) + s(X_t)_-]1_{K_i}(X_t)dA_t]$$

$$= U(a|K_i)(x_o).$$

Here we used the fact that measures η_i are concentrated on K_i.

Finally, if we put:

$$\eta = \sum n_i,\ v = U\eta = \sum U\eta_i = U_B,\ \text{say, we get:}$$

$$\liminf_{\substack{y\to x_o \\ y\in K-\{x_o\}}} (U\eta)(y) = \infty,$$

with

$$E^{x_o}[B^2_\infty]^{1/2} \leq \sum_i [E^{x_o}(A^i_\infty)^2]^{1/2} \leq \sum_i [U(a|_{K_i})(x_o)]^{1/2} < +\infty.$$

Thus, we have proved

THEOREM. *If* K *is thin at* x_o, *there exists a measure* η *with compact support such that: if* $v = U\eta$, *then*

$$\liminf_{\substack{y \to x_o \\ y \in K - \{x_o\}}} v(y) = +\infty,$$

and

$$E^{x_o}[(\sup_t v(X_t))^2] < +\infty.$$

ACKNOWLEDGMENT. The author wishes to express his profound gratitude to Murali Rao for his valuable suggestions concerning this paper.

References

[1] Blumenthal, R. M., and Getoor, R. K., *Markov processes and their potential theory*, Academic Press, New York (1968).

[2] Brelot, M., *On Topologies and Boudaries in Potential theory*, Springer-Verlag, Lecture Notes in Mathematics, no. 175, (1971).

[3] Doob, J. L., *Classical Potential Theory and its Probabilistic Counterpart*, Springer-Verlag, New York, Berlin, Heidelberg, Tokyo, (1984).

[4] Ito, K., and McKean, H. P., Jr., *Diffusion Processes and their sample Paths*, Springer-Verlag, New York, Berlin, Heidelberg (1965).

[5] Port, S., and Stone, C., *Brownian Motion and classical Potential Theory*, Academic Press, New York (1978).

[6] Rao, M. K., Brownian Motion and Classical Potential Theory, Aarhus University, Lecture Notes Series, No. 47, (1977).

[7] Wiener, N., The Dirichlet Problem, J. Math. Phys. 3, 127-146 (1924).

Z. R. Pop-Stojanovic
Department of Mathematics
University of Florida
Gainesville, Florida 32611

Seminar on Stochastic Processes, 1985
Birkhäuser, Boston, 1986

NOTE ON THE GENERATOR OF A RAY RESOLVENT

by

J. Steffens

0. Introduction

The subject of this note is a characterization for the strong generator of a Ray resolvent. The discussion arose from the question if there exists any Hille-Yoshida type theorem in the context of Ray resolvents. More precisely the problem is the following: Given a compact space K and a linear operator L which maps a subspace $\mathscr{D}$ of the space $\mathscr{C}$ of continuous functions on K into $\mathscr{C}$, find necessary and sufficient conditions on $(L, \mathscr{D})$ which ensure that L is the strong generator of a Ray resolvent on K.

Of course the Hille-Yoshida theorem for Banach spaces applied to the uniform closure $\overline{\mathscr{D}}$ of $\mathscr{D}$ yields conditions for the existence of a resolvent of operators on $\overline{\mathscr{D}}$. The remaining problem is to find some additional condition which ensures that the operators extend to kernels on K (that moreover form a Ray resolvent). This requires essentially that functions in $\mathscr{C}$ can be approximated by functions in $\overline{\mathscr{D}}$ in some appropriate sense. In fact, as it turns out, pointwise dominated convergence on what should become the set D of non-branch points is appropriate.

Now in order to obtain a characterization of this set D for a Ray resolvent in terms of its generator we restrict our considerations to Ray resolvents without degenerate branch points and where the constant 1 is excessive. Then the set D coincides with the Choquet boundary $\partial_{\mathcal{D}}K$ of K with respect to the domain $\mathcal{D}$ of the generator. This observation provides the possibility to formulate the desired approximation property, and one obtains the following result:

Assume $1 \in \overline{\mathcal{D}}$ and that $\mathcal{D}$ separates K. Then L is the strong generator of a Ray resolvent if and only if

(i) the operators $\alpha I - L$ map $\mathcal{D}$ onto $\overline{\mathcal{D}}$;

(ii) L satisfies a maximum principle;

(iii) for $f \in \mathcal{C}^+$ there exist $f_n \in \overline{\mathcal{D}}^+$ such that (f_n) converges to f pointwise dominated on $\partial_{\mathcal{D}}K$, and $f_n = (\alpha I - L)g_n$ for an increasing sequence (g_n) in $\mathcal{D}$ and fixed $\alpha > 0$.

The result seems to be of merely theoretical interest (which - to the best of my knowledge - is the case also with the corresponding classical Hille-Yoshida theorem); but it gives e.g. some idea what subspaces of continuous functions appear as domains of generators.

I thank K. Janßen for helpful discussions.

1. Preliminary remarks

Let K be a compact metric space and $(U_\alpha)_{\alpha>0}$ a **Ray resolvent** on K (in the sense of [4]). That means (U_α) is a submarkovian resolvent of kernels on K which maps the set $\mathscr{C}$ of **continuous functions** on K into itself, and for which the set $\mathscr{S}^\beta$ of **continuous β-supermedian functions** with respect to (U_α) separates K for some (hence for all) ($\beta > 0$). Let D denote the set of **non-branch points** of (U_α), $(P_t)_{t>0}$ the associated **Ray semigroup** on K, and $(L, \mathscr{D}_L)$ the **strong generator** of (U_α) or (P_t); that is $\mathscr{D}_L$ denotes the class of those functions $f \in \mathscr{C}$ for which the uniform limit $\underset{t\downarrow 0}{\text{s-lim}}\ \frac{1}{t}(P_t f - f)$ exists in $\mathscr{C}$, and Lf denotes this limit.

Now consider the space β of those functions $f \in \mathscr{C}$ that satisfy $\underset{\alpha\to\infty}{\text{s-lim}}\ \alpha U_\alpha f = f$. Then $\mathscr{B}$ is a Banach space with respect to the sup-norm, and $\mathscr{B}$ coincides with both the set of $f \in \mathscr{C}$ such that $P_0 f = f$ and the set of $f \in \mathscr{C}$ such that $\underset{t\downarrow 0}{\text{s-lim}}\ P_t f = f$. That means (U_α) (resp. (P_t)) is a **strongly continuous resolvent** (resp. **semigroup**) of operators on $\mathscr{B}$.

It is then well known (cf. e.g. [3]) that $\mathscr{D}_L$ is dense in $\mathscr{B}$ and that (U_α) maps $\mathscr{C}$ into $\mathscr{B}$. Furthermore (I denotes the identity)

(1.1) $\alpha I - L$ maps $\mathscr{D}_L$ onto $\mathscr{B}$ for any $\alpha > 0$;

moreover $\alpha I - L$ is one-one and its inverse is U_α. Thus $(\alpha I - L)U_\alpha$ is the identity on $\mathscr{B}$ and $U_\alpha(\alpha I - L)$ is the identity on $\mathscr{D}_L$. Since for $f \in \mathscr{D}_L$ with $f(x_0) > 0$ one has

$$Lf(x_0) = \lim_{t\downarrow 0} \frac{1}{t}(P_t f(x_0) - f(x_0))$$

$$\leq \lim_{t\downarrow 0} \frac{1}{t} \int P_t(x_0,dy)(f(y) - f(x_0))$$

the operator L satisfies the following **maximum principle:**

(1.2) If $f \in \mathcal{D}_L$ attains its maximum value at $x_0 \in K$ and if $f(x_0) \geq 0$, then $Lf(x_0) \leq 0$.

Continuous functions on K now can be approximated by elements of $\mathcal{B} = \overline{\mathcal{D}}_L$ in the following sense:

(1.3) If $f \in \mathcal{C}^+$ and $\alpha > 0$, then $f_n := nU_{n+\alpha}U_\alpha f$ defines an increasing sequence in $\mathcal{D}_L^+$ such that $(\alpha I - L)f_n \in \mathcal{B}^+$ converges to f pointwise dominated on D.

Note for this that $nU_{n+\alpha}f(x) = (n + \alpha)U_{n+\alpha}f(x) - \alpha U_{n+\alpha}f(x)$ converges to f(x) for all $x \in D$ as n tends to infinity.

We now restrict our considerations to a slightly more special situation, namely we assume moreover:

(1.4) the set $\mathcal{E}^\beta$ of continuous β-excessive functions separates K for some (hence for all) $\beta > 0$;

(1.5) 1 is excessive.

This amounts to saying that there are no degenerate branch points and that $P_0 1 = 1$; as a consequence the space $\mathcal{B}$

contains the constants and separates K. Then it is easily verified (cf. e.g. (1.1) and (1.2) of [6]) that

(1.6) the Choquet boundary $\partial_{\mathscr{B}}K$ of K with respect to $\mathscr{B}$ coincides with D, and moreover for any finite measure μ on K the unique equivalent boundary measure is μP_0.

Recall that the **Choquet boundary** $\partial_{\mathscr{B}}K$ consists of all points $x \in K$ where there exists no measure other than ε_x on K that is equivalent with ε_x. Here we call two measures μ, ν on K **equivalent** (with respect to $\mathscr{B}$) if $\mu(f) = \nu(f)$ for all $f \in \mathscr{B}$. Furthermore **boundary measure** refers to a measure carried by $\partial_{\mathscr{B}}K$. Since $\mathscr{D}_L$ is uniformly dense in $\mathscr{B}$ we have of course

(1.7) $\partial_{\mathscr{D}_L}K = \partial_{\mathscr{B}}K = D.$

2. Main result

In combining the properties (1.1), (1.2), (1.3) of Ray resolvents with (1.6), respectively (1.7), one obtains the subsequent characterization for the strong generator. The notations are the same as in section 1, in particular K is a compact metric space and $\mathscr{C}$ denotes the set of continuous functions on K. Furthermore the Choquet boundary of K with respect to some space $\mathscr{D}$ of functions is denoted by $\partial_{\mathscr{D}}K$.

(2.1) THEOREM. *Let* $\mathscr{D}$ *be a linear subspace of* $\mathscr{C}$, *let* $L : \mathscr{D} \to \mathscr{C}$ *be a linear operator, and assume* $\mathscr{D}$ *separates* K *and* $1 \in \overline{\mathscr{D}}$ *(the uniform closure of* $\mathscr{D}$ *in* $\mathscr{C}$*). Then* $(L, \mathscr{D})$ *is the strong generator of a Ray resolvent on* K *satisfying* (1.4) *and* (1.5) *if and only if the following conditions hold:*

(i) $\alpha I - L$ *maps* $\mathscr{D}$ *onto* $\overline{\mathscr{D}}$ *for any* $\alpha > 0$;

(ii) *if* $f \in \mathscr{D}$ *attains its maximum value at* $x_0 \in K$ *and if* $f(x_0) \geq 0$, *then* $Kf(x_0) \leq 0$;

(iii) *there exists* $\alpha > 0$ *such that for any* $f \in \mathscr{C}^+$ *there exists a sequence* (f_n) *in* $\overline{\mathscr{D}}^+$ *that converges to* f *pointwise dominated on* $\partial_{\mathscr{D}} K$, *and* $f_n = (\alpha I - L)g_n$ *for some increasing sequence* (g_n) *in* $\mathscr{D}$.

PROOF. The necessity of these conditions was explained in the previous section. As for the sufficiency, note first that the maximum principle for L in (ii) implies $\|\alpha f - Lf\| \geq \alpha\|f\|$ for $f \in \mathscr{D}$ and $\alpha > 0$, because $\|(\alpha I - L)f\| \geq (\alpha I - L)f(x_0) \geq \alpha f(x_0) = \alpha\|f\|$ if $f(x_0) \geq 0$ is the maximum value (otherwise the same argument works with $-f$).

Hence by the version in [3] of the Hille-Yoshida theorem applied to the Banach space $\overline{\mathscr{D}}$, there exists a resolvent $(U_\alpha)_{\alpha>0}$ of operators U_α on $\overline{\mathscr{D}}$ satisfying $\|U_\alpha\| \leq \frac{1}{\alpha}$ which are the inverse operators of $\alpha I - L$. Moreover, by using (ii) again one shows that U_α is a positive operator for any $\alpha > 0$, in fact: let $g \in \overline{\mathscr{D}}$, $g \geq 0$, then if $f := U_\alpha g$ attains its minimum value at x_0 one obtains

$$f > f(x_0) > f(x_0) - \frac{1}{\alpha}Lf(x_0) = \frac{1}{\alpha}(\alpha I - L)f(x_0) > 0.$$

Therefore $f \to \alpha U_\alpha f(x)$ defines a positive real valued mapping on $\overline{\mathscr{D}}$ (for $\alpha > 0$ and $x \in K$), which by the Hahn-Banach theorem can be extended to a linear functional on all of $\mathscr{C}$ that moreover is positive and thus defines a measure μ_x^α of K. (As for the positivity observe that $\mathscr{C} = \mathscr{C}^+ + \overline{\mathscr{D}}$ such that e.g. th. 34.2 of [2] applies.)

Now for $\alpha > 0$ and $x \in K$, there exists a boundary measure ν_x^α on K equivalent with μ_x^α with respect to $\overline{\mathscr{D}}$ (see e.g. [1] or the remarks in section 1 of [6]); i.e. ν_x^α is carried by $\partial_{\mathscr{D}} K = \partial_{\overline{\mathscr{D}}} K$, and

$$\frac{1}{\alpha}\nu_x^\alpha(f) = \frac{1}{\alpha}\mu_x^\alpha(f) = U_\alpha f(x)$$

for any $f \in \overline{\mathscr{D}}$. Since for $f \in \mathscr{C}^+$ and $\beta > 0$ we have

$$\nu_x^\beta(f) = \lim_n \nu_x^\beta(f_n) = \lim_n U_\beta f_n(x),$$

where (f_n) is chosen according to (iii), any $\nu_\cdot^\beta$ defines a kernel on K, which we again denote by U_β.

We claim that these kernels $(U_\beta)_{\beta>0}$ form a Ray resolvent on K. At first, the resolvent equation carries over from $\overline{\mathscr{D}}$ to $\mathscr{C}$ by means of condition (iii) as above. To show the remaining properties we need the second part of (iii); let therefore $\alpha > 0$ be fixed, let $f \in \mathscr{C}^+$, and $f_n = (\alpha I - L)g_n$ be chosen as in (iii). Then

$$U_\alpha f = \lim_n U_\alpha f_n = \lim_n g_n = \sup_n g_n$$

is lower semicontinuous, hence continuous since $1 \in \overline{\mathscr{D}}$ and so (for $0 \leq f \leq 1$): $U_\alpha f + U_\alpha(1 - f) = U_\alpha 1 \in \mathscr{C}$. Consequently U_α maps $\mathscr{C}$ into itself, but since $\sup_n g_n = U_\alpha f \in \mathscr{C}$, by Dini's lemma the convergence is uniform and thus $U_\alpha f \in \overline{\mathscr{D}}$. This now implies that for any $\beta > 0$ and $f \in \mathscr{C}$ one has $U_\beta f = U_\alpha f + (\alpha - \beta)U_\beta U_\alpha f \in \overline{\mathscr{D}}$. Hence the resolvent $(U_\beta)_{\beta>0}$ maps $\mathscr{C}$ into itself.

Finally, since $\mathscr{D}$ separates K and $\mathscr{D} = U_\alpha(\overline{\mathscr{D}})$, we obtain in particular that the set $\mathscr{E}^\alpha (\subset \mathscr{S}^\alpha)$ of continuous α-excessive functions with respect to (U_β) separates K, which completes the proof. □

(2.2) REMARK. If in particular the domain $\mathscr{D}$ of L is dense in $\mathscr{C}$ the classical Hille-Yoshida theorem is obtained, since then condition (iii) follows trivially from (i) by $f_n := f$.

(2.3) REMARKS. a) It suffices to require condition (i) only for some $\alpha > 0$, because then the solvability of the equations can be derived as usual for $0 < \beta < 2\alpha$, and so on (cf. e.g. [5]).

b) For the maximum principle in (ii) only points $x_0 \in \partial_{\mathscr{D}} K$ have to be considered, since any function in $\mathscr{D}$ attains its maximum value at a point of the Choquet boundary.

c) The proof of (2.1) shows that in order to obtain a resolvent of kernels on K (without further smoothness properties etc.), besides conditions (i) and (ii) only the first part of condition (iii) is needed, namely: for any $f \in \mathscr{C}^+$ there exists a sequence (f_n) in $\overline{\mathscr{D}}^+$ that converges to f pointwise dominated on $\partial_{\mathscr{D}}K$.

d) The restrictions on $\mathscr{D}$ in the statement of (2.1) are not crucial, but simplify the presentation. The assumption that $\mathscr{D}$ separates K can be removed; if one considers the quotient space $\tilde{K}$ generated by identifying those points that cannot be separated by $\mathscr{D}$, one can construct the resolvent on $\tilde{K}$ and then extend it to K, accordingly. The assumption $1 \in \overline{\mathscr{D}}$ can be weakened to: there exists a strictly positive $p \in \overline{\mathscr{D}}$. Then with some appropriate changes in the conditions (i)-(iii), one obtains a more general version of theorem (2.1), which by means of a Doob transformation reduces to the situation in (2.1).

e) It is not immediately clear if the conditions can be weakened by requiring e.g. a densely defined range for $\alpha I - L$ in (i). The usual considerations on closability of the operator L etc. (as in [5]) depend heavily on the denseness of the domain of L in $\mathscr{C}$.

References

[1] Alfsen, E. M.: Compact Convex Sets and Boundary Integrals. Berlin-Heidelberg-New York Springer 1971.

[2] Choquet, G.: Lectures on Analysis, Vol. II. New York-Amsterdam W. A. Benjamim Inc. 1969.

[3] Dynkin, E. B.: Markov Processes, Vol. I. Berlin-Heidelberg-New York Springer 1965.

[4] Getoor, R. K.: Markov Processes: Ray Processes and Right Processes. Lecture Notes in Math. 440 Berlin-Heidelberg-New York Springer 1975.

[5] Sato, K.: Semigroups and Markov Processes. Lectures at the University of Minnesota 1968.

[6] Steffens, J.: A Sheaf Property for Excessive Functions of Right Processes. To appear in Math. Nachr.

Jutta Steffens
Institut für Statistik und Dok.
Universität Düsseldorf
Universitätsstr. 1
D-4000 Düsseldorf 1
W. - Germany

Seminar on Stochastic Processes, 1985
Birkhäuser, Boston, 1986

INFINITE EXCESSIVE AND INVARIANT MEASURES*

by

Michael I. Taksar

1. Formulation of Results

1.1. In the paper [10] the following problem was considered. Given a contraction semigroup T_t on a Borel space D and an excessive measure ν, when is it possible to find another contraction semigroup $\tilde{T}_t$ such that

(1.1.1) $\tilde{T}_t \geq T_t$

and

(1.1.2) $\nu\tilde{T}_t = \nu.$

* This research was sponsored by Office of Naval Research Contract No. 00014-79-C-0685 at the Institute for Mathematical Studies in the Social Sciences, Stanford University.

The most restrictive condition under which this problem was solved is the finiteness of the excessive measure ν. This condition excludes such an interesting case as the semigroup T_t generated by the transition function of Wiener's process killed at the origin and the Lebesque measure ν.

If ν is a finite measure then the semigroup $\tilde{T}_t$ is conservative, i.e.,

1.1.A. For all $t > 0$, $\tilde{T}_t 1(x) = 1$, for all $x \in D$.

More precisely, any $\tilde{T}_t$ subject to (1.1.1) - (1.1.2) satisfies 1.1.A for ν almost all x, and it is always possible to find such $\tilde{T}_t$ for which 1.1.A holds for all $x \in D$. Unfortunately, if ν is an infinite measure, then $\tilde{T}_t$, when it exists, is not necessarily a conservative semigroup, as the examples in Section 1.3 show.

Our main goal is to prove that one still can find a conservative semigroup $\tilde{T}_t$ satisfying (1.1.1) - (1.1.2), if ν is a quasi-finite null-excessive measure.

DEFINITION Let T_t be a semigroup. An excessive measure ν is called quasi-finite with respect to T_t if for some $s > 0$ the difference between ν and νT_s is a finite measure.

A measure ν is called null-excessive with respect to T_t if for each $\Gamma \subset D$, subject to $\nu(\Gamma) < \infty$

$$\nu T_t(\Gamma) \downarrow 0 \quad \text{as} \quad t \to \infty.$$

The principal part of the proof of the main result is the same as that of [10]. It is done in Section 3. We consider the transition function p which generates T_t, then we construct a stationary Markov process $(w(s),P)$ with the transition function p and the one-dimensional distribution ν. (Actually, the process $w(\cdot)$ has random birth and death times and the measure P is infinite.) We add a single point V to the space D and we look for a stationary Markov process $(x_t,\overline{P})$ with the state space $E = D \cup V$ such that

1.1.α. The birth time of x_t is equal to $-\infty$ and the death time of x_t is equal to $+\infty$.

1.1.β. The one-dimensional distribution of $\overline{P}$ is equal to ν.

1.1.γ. $p(t,x;\Gamma) = \overline{P}_x\{x_t \in \Gamma;\ x_s \overline{\in} V \text{ for all } s < t\}$.

A process $(x_t,\overline{P})$ satisfying 1.1.α - 1.1.γ is called a covering process for $(w(s),P)$, and $(w(s),P)$ is called a subprocess of $(x_t,\overline{P})$. The transition function of $(x_t,\overline{P})$ generates the conservative semigroup $\tilde{T}_t$ we are looking for.

It is important to mention that we do not suppose the process $(w(s),P)$ to have any regularity properties at all. As a result, the covering process $(x_t,\overline{P})$ is not regular, but has some kind of regular behavior at the first hitting times of V; that is, these times are measurable random variables and some analog of a strong Markov property for these times is true (see Lemma 3.3.4). As a consequence we do not need any regularity conditions for T_t

and $\tilde{T}_t$ in the formulations of the main theorems.

The construction of a covering process for (w(s),P) is based on the theory of translation invariant regenerative sets on real line, developed in Section 2. Any set of this kind forms a probabilistic replica of itself after each stopping time t which belongs to this set, and the probability distribution of such a set is invariant under shift operations. We will show that if we permit infinite underlying probability distribution on the sample space, then all such sets are in one-to-one correspondence with ranges of all processes with independent increments. (If we restrict ourselves to finite underlying probability measures, then we should consider only those processes with independent increments whose mean is finite. This case was treated in [8]). Here the most important tool is the theorem of B. Maisonneuve in [7], which enables us to find an invariant distribution for the "jump process" of the process with independent increments.

In the second part of the section we give a precise formulation of the main results and give the conditions under which they are proved.

Section 1.3 is devoted to counterexamples. Here we explain the probabilistic meaning of the conditions involved in the formulation of the main result, and why these conditions cannot be dropped.

As always, the same letter is used for a measure and the integral with respect to this measure. The word "function" stands for a nonnegative bounded measurable function.

In case when the proof is similar to the one given in

[8], [9] or [10], we shall only outline it without going into details.

1.2. Let D be a Borel space and T_t, $t > 0$, be a linear semigroup in the Banach space of bounded measurable functions on D (we say for brevity that T_t is a semigroup on D). We always assume $T_t f(x)$ to be measurable in (t,x). The semigroup T_t is called a positivity preserving normal contraction semigroup if

1.2.A. For any $t > 0$ and each function $g > 0$ $T_t g > 0$.

1.2.B. For each $x \in D$

$$T_t 1(x) < 1, \quad \text{and} \lim_{t \downarrow 0} T_t 1(x) = 1.$$

1.2.C. If $f(x_0) = 0$ then $T_0 f(x_0) = 0$.

A semigroup T_t is called continuous if

1.2.D. For each $x \in D$ $T_t 1(x)$ is a continuous function of $t > 0$.

(Note that 1.1.A implies 1.2.D). A positivity preserving normal contraction semigroup is denoted S-semigroup.

If S-semigroup T_t satisfies 1.2.E below, then T_t is called dying or SD-semigroup; if T_t satisfies 1.1.A, then T_t is called conservative or SC-semigroup.

1.2.E. For each $x \in D$ $\lim_{t \to \infty} T_t 1(x) = 0$.

If T_t and $\tilde{T}_t$ are two semigroups on D and for each positive function g

$$T_t g \le \tilde{T}_t g, \tag{1.2.1}$$

then we say that $\tilde{T}_t$ is larger than T_t, or $\tilde{T}_t$ is an enhancing of T_t.

We write $T_t = \tilde{T}_t$ a.e.μ if for any function g for μ-almost all x, $T_t g(x) = \tilde{T}_t g(x)$.

In this paper we are going to prove the following theorems.

THEOREM 1 Given a continuous SD-semigroup T_t and a quasi-finite null excessive measure ν, one can find a SC-semigroup $\tilde{T}_t$ which is larger than T_t and for which ν is invariant.

THEOREM 2 If T_t and ν satisfy the conditions of Theorem 1 and if in addition ν is an extreme excessive measure then $\tilde{T}_t$ is unique up to the measure ν.

1.3. In this section we explain the probabilistic meaning of the conditions in the formulations of the main theorems and give examples which show that these conditions cannot be dropped. All the examples are either derived from deterministic movement or diffusion. Here and in the sequel T denotes the real line $]-\infty,+\infty[$ and T_+ the positive half line $[0,+\infty[$.

First consider quasi-finiteness. Let p be the transition function which generates T_t. Consider a process w(s), s > 0 with initial distribution ν and transition function p. Then

$$(1.3.1) \qquad P\{\beta < u\} = (\nu - \nu T_u)(D),$$

where β is the time of death of w(s). In Section 3 we show that (1.3.1) is finite for all u. Hence, quasi-finiteness can be formulated as an existence of a distribution function for the death time β. In other words, the "projection" of measure P on the "β-axis" is σ-finite.

Consider a semigroup on the real line

$$T_t f(x) = e^{-\alpha t} f(x + t).$$

Let ν be a measure with density $\lambda(x) = \exp(x)$. This measure is null-excessive but not quasi-finite. The semigroup T_t satisfies all the conditions of Theorem 1. The semigroup $\tilde{T}_t$ which preserves ν is given by

$$\tilde{T}_t f(x) = e^t T_t f(x + t)$$

which is not conservative.

Now let T_t be semigroup generated by the transition function of a Brownian motion on T_+ killed at the origin. Let ν have a density $\lambda(x) = x$. Elementary calculations show that ν is an invariant measure. Measure ν is a quasi-finite (but not null-excessive) and T_t is a dying continuous semigroup. Since ν is invariant, the only semigroup which preserves ν and is larger than T_t must coincide with T_t, a.e., ν. Hence, no conservative semigroup of such kind exists.

The next example shows that in the formulation of Theorem 1 the condition 1.2.E cannot be dropped. Put

$$T_t f(x) = f(x + t), \quad x \in T.$$

Let ν have density

$$\lambda(x) = \begin{cases} 2 - e^{-x}, & x > 0 \\ e^{x}, & x < 0. \end{cases}$$

It is obvious that ν is excessive and $\nu - \nu T_t$ is a finite measure for any $t > 0$. Nevertheless, any conservative $\tilde{T}_t$ which is larger than T_t must coincide with T_t and, therefore, cannot satisfy Theorem 1.

Consider ν as the initial distribution of masses (particles) on D, rather than the probability, and p as the law which governs the redistribution of masses in time. The condition 1.2.E together with null excessivity says that any subset of D which contains finite mass eventually loses everything, and this is due entirely to the death of the particles, rather than to their transient behavior.

The following example shows that we cannot drop the condition 1.2.D. In terms of stochastic processes, this condition means that, given any initial state x   D, the distribution of β does not have atoms. The continuity of T_t is required both in the cases of finite and infinite measure ν.

Let D be an interval $]0,1[$, ν be the uniform distribution on D, and

$$T_t f(x) = f(x + t)1_{\{x+t<1\}}, \quad x \in D.$$

Any enhancing $\tilde{T}_t$ of T_t is described by a distribution function F on $]0,1[$ ($F(0+) = 0$, $F(1-) = 1$), such that

$$\tilde{T}_{1-x} f(x) = \int_0^1 f(x)dF(x).$$

If we consider a renewal process with underlying distribution F, then ν must be a stationary renewal measure for this process. Such a measure is unique and has density

$$g(x) = 1 - F(x) \tag{1.3.2}$$

(see [5], Ch. 5). Formula (1.3.2) implies that $F(x) = 0$ for all $x \in]0,1[$, and it contradicts the fact that F is a distribution function of a probability measure on $]0,1[$.

REMARK The same arguments work for any homogenuous Markov process with deterministic death, e.g., space-time Brownian Bridge.

It should be mentioned that we can drop the requirement that T_t is continuous, if we permit $\tilde{T}_t$ to be a semigroup on a larger space $E \supset D$ (in the example above $E =]0,1]$). This is true for finite as well as for infinite measures ν.

2. Regenerative Sets with Infinite Underlying Measures

2.1. Let $(\Omega, \mathscr{F})$ be a measurable space and Q be a measure on $\mathscr{F}$ (not necessarily finite). A subset M $T \times \Omega$ is called a random set (r.s.) if it is $\mathscr{B} \times \mathscr{F}$-measurable and $M(\omega)$ is nonempty for a.e., ω. (Here T is the real line $]-\infty,+\infty[$ and $\mathscr{B}$ is its Borel σ-field.) A r.s. M is called closed (closed from the right, perfect, discrete, etc.) if for a.e. ω, $M(\omega)$ is closed (closed from the right, perfect, discrete, etc). Only closed random sets will be considered in the sequel.

With any r.s. M a two-dimensional stochastic process $z_t = (z_t^-, z_t^+)$ is associated:

$$z_t^+(\omega) = \inf \{s > t : s \in M(\omega)\}$$

$$z_t^-(\omega) = \sup \{s < t : s \in M(\omega)\}.$$

The distribution of a r.s. means the distribution of its associated process, as well as the independence or conditional independence of two random sets means independence or conditional independence of their associated process. We denote by M_t (by M^t) the intersection of M with $]-\infty,t]$ (with $[t,+\infty[$) and we put (for a set M and a number y we denote by $M - y$ the set $\{x - y : x \in M\}$)

$$\tilde{M}_t(\omega) = M_t(\omega) - z_t^-(\omega)$$

$$\tilde{M}^t(\omega) = M^t(\omega) - z_t^+(\omega).$$

The set M is called right regenerative (r.r.) if

2.1.α. M_t and $\tilde{M}^t$ are conditionally independent, given z_t^+, and the distribution of $\tilde{M}^t$ does not depend on t.

The set M is called left regenerative (ℓ.r.) if

2.1.β: M^t and $\tilde{M}_t$ are conditionally independent, given z_t^-, and the distribution of $\tilde{M}_t$ does not depend on t.

The set, satisfying both 2.1.α and 2.1.β, is called regenerative (r.).

The set M is translation invariant (t.i.) if

2.1.γ. The distribution of M - t is independent of t. This is equivalent to the stationarity of the process $z_t - (t,t)$.

Increasing processes with independent increments serve as a very powerful tool in studying r. sets.

Let α be a nonnegative constant and Π be a measure on $]0,\infty[$ such that

$$(2.1.1) \qquad \int_0^\infty x \wedge 1 \; \Pi(dx) < \infty.$$

Then there exists a right-continuous increasing process y_t with independent increments (subordinator), $t \in T_+$, with transition probabilities Q_ℓ and the set of discontinuities J such that

For any function f on T

$$(2.1.2)\quad Q_\ell\{\sum_{t\in J, c<t\leq d} f(y_t - y_{t-})\} = (d - c)\int_0^\infty f(x)\Pi(dx)$$

$$(2.1.3)\quad y_d - y_c = \alpha(d - c) + \sum_{t\in J, c<t\leq d}(y_t - y_{t-}).$$

We call y_t an (α,Π)-process. The constant α is called the translation constant and Π is called the Levy measure of the process.

Any two (α,Π)-processes differ only by initial distributions. The range, - i.e., the closure of the set of values - of any (α,Π)-process is a r.r. set (see [8], Sec. 6).

A r.s. M is called (α,Π)-generated, if for each t, M^t has the same distribution as the range of an (α,Π)-process. Note, that any (α,Π)-generated set is r.r.

In [8], all r.r.t.i. sets with finite underlying measure Q were studied. All sets of such kind are (α,Π)-generated, with Π having finite first moment. Our goal is to extend the same results to infinite measures Q. A r.s. M is said to have a σ-finite distribution (or M is a σf-set) if

2.1.A. The process z_t has σ-finite one-dimensional distributions.

For example, consider any increasing process with independent increments with the Lebesque initial distribution (i.e., initial distribution uniform on T). The range of this process is a r.s. whose one-dimensional distributions are not σ-finite. Let us take now any σ-

finite measure ν with support on $[0,1]$ and let Π be a unit measure concentrated at the point 1. If we consider the range of the $(0,\Pi)$-process with initial distribution ν, then this r.s. has a σ-finite distribution.

It is possible to give a complete analysis of r.r.t.i. σf-sets, similar to the one given in [8], but we restrict ourselves only to the theorem of existence.

THEOREM 2.1.1 *For any measure* Π *on* $]0,\infty[$ *subject to* (2.1.1) *there exists a* t.i. $(0,\Pi)$-*generated* σf-*set* M. *The set* M *is left regenerative and moreover,* $-M$ *has the same distribution as* M.

Let the complement of M be the union of disjoint open intervals $]\gamma,\delta[$. Then for any function f on $T \times T$,

$$Q\{\sum_{\gamma} f(\gamma,\delta)\} = \int_{-\infty}^{\infty}\{\int_{0}^{\infty} f(s,s+y)\Pi(dy)\}ds. \tag{2.1.4}$$

2.2. Let y_t be a $(0,\Pi)$-process and Q_x be its transition probabilities. We denote by σ_ℓ the first hitting time of $]\ell,\infty[$ by y_t; and by $Y_\ell = (U_\ell,V_\ell) \equiv (y_{\sigma_\ell -}, y_{\sigma_\ell})$ we denote the "jump" process of y_t. Due to the strong Markov property of y_t, V_t as well as Y_t is a Markov process. Let

$$q(s,x;\ t,\Gamma) = \begin{cases} 1_\Gamma(x) & , \text{ if } x > t \\ Q_x\{V_t \in \Gamma\}, & \text{if } x < t,\ \Gamma \subset T \end{cases}$$

be the transition function of the process V_t. Let

$$\Pi(x;\Gamma) = \Pi(\Gamma - x),\quad \Gamma \subset T,\ x \in T;$$

$$R_t =]-\infty, t[, \quad R^t =]t, \infty[$$

$$\mu_t(\Gamma) = \int_{-\infty}^{t} \Pi(x; \Gamma)dx, \quad \Gamma \subset R^t$$

By the theorem of Maisonneuve (see [7], Th. (3.2)) the family μ_t is an entrance law with respect to q. Note that $\mu_t(\Gamma) = \mu_0(\Gamma - t)$. Consider the Markov process (v_t, Q) with the one-dimensional distributions μ_t and with the transition function q. (The measure Q is finite iff μ_0 is a finite measure.) The existence of such a Markov process is proved in [6]. The same way as in Lemma 6.2 of [8], we can show that v_t is a stochastically continuous increasing process; hence, there exists a right-continuous version of it. Consider the random set M which is the range of v_t (i.e., the closure of the set of values of v_t). We are going to prove that M is the set we are looking for.

Let

$$\Pi_x(\Gamma \times \Delta) = 1_x(\Gamma)\Pi(x; \Delta), \quad \Gamma, \quad \Delta \subset T, \quad x \in T.$$

$$\lambda_b(\Gamma) = Q_b\{\int_0^{\infty} 1_\Gamma(y_t)dt\}, \quad x \in T, \quad \Gamma \subset T$$

$$\Pi^*(x; \Gamma) = \Pi(-x; -\Gamma)$$

$$\lambda_b^*(\Gamma) = \lambda_{-b}(-\Gamma).$$

LEMMA 2.2.1: <u>For any function</u> f <u>on</u> $T \times T$

$$Q_b\{\sum_{t \in J} f(y_{t-}, y_t)\} = \int_T \lambda_b(dx)\Pi_x(f). \tag{2.2.1}$$

The proof of this lemma is well known.

LEMMA 2.2.2: *If* f *and* g *are the functions on* R_s *and* R^s *respectively, then*

$$\int_0^{\infty} \lambda_x(g)f(x)dx = \int_0^{\infty} g(x)\lambda_x^*(f)dx \tag{2.2.2}$$

$$\int_{-\infty}^{s} \Pi(x;g)f(x)dx = \int_s^{\infty} \Pi^*(x;f)g(x)dx. \tag{2.2.3}$$

For the proof see [8, L6.4].

LEMMA 2.2.3 *The set* M *is a translation invariant right-regenerative* $(0,\Pi)$-*generated set with the associated process* z_t *having the one-dimensional distributions*

$$\nu_t(\Gamma) = \int_{-\infty}^{t} \Pi_x(\Gamma)dx, \quad \Gamma \subset R_t \times R^t. \tag{2.2.4}$$

PROOF *1°*. Fix $s \in T$. Consider a $(0,\Pi)$-process $y_\cdot$ with initial distribution μ_s. Let $V_u = y_{\sigma_u}$. By the construction of (v_t, Q) the process v_t, $t > s$ has the same finite-dimensional distributions as V_t, $t > s$. Both processes are right-continuous, therefore, their ranges have equal distributions. But the range of $V_\cdot$ is equal to that of $y_\cdot$, and that proves that M is $(0,\Pi)$-generated (right-regenerativity is a consequence of this fact).

By the construction, the process $v_t - t$ is Markov with stationary transition function and stationary one-dimensional distributions (equal to μ_0). Hence, M is a t.i. set.

2°. For any $(0,\Pi)$-process,

$$(2.2.5) \qquad y_t = y_0 + \sum_{s \in J} (y_s - y_{s-}).$$

Formula (2.2.5) implies that the range of any $(0,\Pi)$-process has Lebesque measure zero, a.s., Q. Since M is a t.i. set, and M^t is the range of $(0,\Pi)$-process,

$$(2.2.6) \qquad Q\{t \in M\} = \int_t^{t+1} Q\{s \quad M\}ds = Q\{\int_0^1 1_M(s)ds\} = 0.$$

<u>3°</u>. To prove (2.2.4) we can consider only bounded sets Γ of the form $\Delta_1 \times \Delta_2$. By virtue of (2.2.6), $Q\{z_t = (t,t)\} = 0$; consequently we may take $\Delta_1 < t$ and $\Delta_2 > t$. Since Δ_1 is bounded there exists s such that $\Delta_1 > s$.

Since M is $(0,\Pi)$-generated

$$Q\{z_t \in \Gamma | z_s^+\} = Q_{z_s^+}\{Y_t \in \Gamma\},$$

a.e., Q on the set $\{z_s^+ < t\}$, $\Gamma \in R_t \times R^t$.

The distribution of z_s^+ is equal to that of v_s, and we can write

$$(2.2.7) \qquad \begin{aligned} Q\{z_t \in \Gamma\} &= \int_{R^s} \mu_s(dx) Q\{z_t \in \Gamma | z_s^+ = x\} \\ &= \int_{R^s} \mu_s(dx) Q_x\{Y_t \in \Gamma\}. \end{aligned}$$

The last equality in (2.2.7) due to the fact that $\{z_s^+ < t\} = \{z_t^- > s\}$. By virtue of Lemma 2.2.1 the right hand side of (2.2.7) is equal to

$$(2.2.8)\qquad \int_{-\infty}^{s} dy \int_{s}^{t} \Pi\{y;\ dx\} \int_{\Delta_1} \lambda_x(dz)\Pi(z;\ \Delta_2).$$

Applying (2.2.3) to (2.2.8), we get

$$(2.2.9)\qquad \int_{s}^{t} dy\Pi^*(y;\ R_s) \int_{\Delta_1} \lambda_y(dz)\Pi(z;\ \Delta_2).$$

Applying (2.2.2) to (2.2.9), we get

$$(2.2.10)\qquad \int_{\Delta_1} dx\Pi(x;\ \Delta_2) \int_{s}^{t} \lambda_x^*(dz)\Pi^*(z;\ R_s).$$

(The upper limit in the last integral in (2.2.10) can be changed to z, because the support of the measure λ_z^* is R_z).

Let $y_t^* = -y_t$ be the decreasing process with independent increments. The Levy measure of y_t^* equals to Π^*, where

$$\Pi^*(\Gamma) = \Pi(-\Gamma).$$

By virtue of Lemma 2.2.1, for each $s < z < t$

$$(2.2.11)\qquad \int_{s}^{z} \lambda_z^*(dy)\Pi^*(y;\ R_s) = Q_z^*\{y^*_{\sigma_s^*} < s\},$$

here Q_z^* is the transition probability of $y^*_{\bullet}$ and σ_s^* is the first hitting time of $]s,-\infty[$. Since for each a

$$f(z,s,a) = Q^*_{z+a}\{y^*_{\sigma^*_{s+a}} = s + a\}$$

does not depend on a and the range M^* of a $(0,\Pi^*)$ - process has Lebesque measure zero, we get

$$f(z,s,0) = \int_0^1 f(z,s,a)da = Q_z^*\{\int_s^{s+1} 1_x(M^*)dx\} = 0,$$

therefore, the right-hand side of (2.2.11) differs from 1 only for z of Lebesque measure zero. Hence (2.2.10) equals

(2.2.12) $$\int_{\Delta_1} \Pi(x;\ \Delta_2)dx = \int_{-\infty}^{t} \Pi_x(\Delta_1 \times \Delta_2)dx.$$

LEMMA 2.2.4 <u>The distribution of</u> $-M$ <u>is equal to that of</u> M.

PROOF <u>1°</u>. Let $x,y,s,t \in T$, $z = (x,y)$ and $\Gamma \subset T$. Put

$$p(s,z;\ t,\Gamma) = \begin{cases} 1_\Gamma(z) & \text{if } y > t, \\ Q_y\{Y_t \in \Gamma\} & \text{if } y < t. \end{cases}$$

Here Q_y is the transition probability of $(0,\Pi)$-process $y_\cdot$, and Y_t is the "jump process" of $y_\cdot$, defined at the beginning of this subsection. Since M is $(0,\Pi)$-generated set, the associated process z_t is Markov with the transition function p, given by (2.2.13). Consequently, ν_t defined by (2.2.4), is an entrance law with respect to p.

For $z = (x,y)$, put $\bar{z} = (y,x)$. If $\Gamma \subset T \times T$ then $\bar{\Gamma}$ must be understood in the same way. Let

(2.2.14) $$\nu_t^*(\Gamma) = Q\{-\bar{z}_t \in \Gamma\} = Q\{z_t \in -\bar{\Gamma}\}.$$

(2.2.15) $$p^*(s,z;\ t,\Gamma) = p(-t,-\bar{z};\ -s,-\bar{\Gamma}).$$

Let z_t^* be the associated process of $-M$. It is obvious that $z_t^* = -\bar{z}_{-t}$, where z_t is the associated process of M.

Therefore, ν^*_{-t} is the one-dimensional distribution (at time t) and p^* is the backward transition function of z^*_t. If we could prove that $\nu^*_t = \nu_{-t}$ and p^* is the backward transition function of z_t, then it would follow that M and -M have the same distributions.

<u>2°</u>. Let $\Gamma = \Delta_1 \times \Delta_2$, $\Delta_1 < t$, $\Delta_2 > t$. Applying successfully (2.2.12) and (2.2.3),

$$Q\{z_t \in -\overline{\Gamma}\} = \int_{-\infty}^{t} \Pi_x(-\overline{\Gamma})dx$$

$$= \int_{t}^{\infty} \Pi^*_x(-\Gamma)dx = \int_{-\infty}^{-t} \Pi_x(\Gamma)dx = Q(z_{-t} \in \Gamma).$$

Therefore $\nu_{-t} = \nu^*_t$.

<u>3°</u>. To calculate the backward transition function of z_t, we should consider the two-dimensional distribution of $z_\cdot$.

Let

$$\Gamma = F \times G;\ F,\ G \subset T;\ F < s;\ s < G < t;$$

$$\Delta = H \times K;\ H,\ K \subset T;\ s < H < t;\ H > G;\ K > t.$$

Owing to the fact that ν_t is the one-dimensional distribution and p is the transition function of z_t, we can write (in the last equality below we use (2.2.1))

$$\begin{aligned} &Q\{z_s \in \Gamma,\ z_t \in \Delta\} \\ (2.2.16)\qquad &= \int_\Gamma \nu_s(dz)p(s,z;\ t,\Delta) \\ &= \int_F dx \int_G \Pi\ (x;\ dy) \int_H \lambda_y(du)\Pi(u;\ K). \end{aligned}$$

Applying successfully (2.2.3) and (2.2.2) to (2.2.16), we get that (2.2.16) equals

$$(2.2.17) \qquad \int_H dx\Pi(x;\ K) \int_G \lambda_x^*(du)\Pi^*(u;\ F).$$

Note that p^*, defined by (2.2.15) is the backward transition function of the jump process of y_t^*, i.e.,

$$p^*(t,(x,y);\ s,G \times F) = Q_x^*\{y^*_{\sigma_s^*-} \in G, y^*_{\sigma_s^*} \in F\}.$$

For $G > s$ and $F < s$

$$1_{G\times F}(y^*_{\sigma_s^*-}, y^*_{\sigma_s^*}) = \sum_{t\in J} 1_{G\times\Gamma}(y_{t-}, y_t).$$

Therefore, by virtue of Lemma 2.2.1

$$(2.2.18) \qquad p(t,(x,y);\ s,G \times F) = \int_G \lambda_y^*(du)\Pi^*(u;\ F).$$

Substituting (2.2.18) in (2.2.17), we get

$$(2.2.19) \qquad Q\{z_s \in \Gamma, z_t \in \Delta\} = \int_\Delta \nu_t(dz)p^*(t,z;\ s,\Gamma).$$

The proof of (2.2.19) for general Δ and Γ is similar. Therefore, p^* is the backward transition function of z_t.

This finishes the proof that $-M$ has the same distribution as M and, therefore, M is left regenerative.

LEMMA 2.2.5 <u>The set</u> M <u>satisfies</u> (2.1.4).

PROOF In (2.1.4) we may consider only the functions f such that

$$(2.2.20) \qquad f(x,y) = 0 \quad \text{if } x > y.$$

Put $R_{st} = \{(x,y) : x < s, y < t\}$, $f_{st} = f1_{R_{st}}$. For $\Lambda = r_1, r_2, \ldots, r_k$ set $R_\Lambda = R_{r_1 r_1} \cup R_{r_2 r_2} \cup \ldots \cup R_{r_k r_k}$, $f_\Lambda = f1_R$. If $r_1, r_2, \ldots, r_k$ is a sequence of all rational numbers and $\Lambda(n) = \{r_1, \ldots, r_n\}$, then $f_{\Lambda(n)} \uparrow f$ for any function f subject to (2.2.20). Trivial computations show that the function $f_{\Lambda(n)}$ is a linear combination of the functions f_{st}, $s < t$. Since both sides of (2.1.4) are stable under linear operations and monotone passage to the limit, we have to verify (2.1.4) only for the functions f_{st}, $s < t$.

$$\begin{aligned} Q\{\sum_\gamma f_{st}(\gamma,\delta)\} &= Q\{\sum_\gamma 1_{\gamma<s} f(\gamma,\delta) 1_{\delta>t}\} \\ &= Q\{f_{st}(z_t)\} \\ &= \nu_t(f_{st}) \\ &= \int_{-\infty}^{t} \Pi_x(f_{st})dx \\ &= \int_{-\infty}^{\infty} \Pi_x(f_{st})dx \\ &= \int_{-\infty}^{\infty} dx \int_0^{\infty} f_{st}(x, x+y)\Pi(dy). \end{aligned}$$

LEMMA 2.2.6 Let Q_y and Q_y^* be the transition probabilities of the $(0,\Pi)$-process y_t and the process

$y_t^* \equiv -y_t$ respectively. For a function F on $(T \times T)^n$ and a function G on $(T \times T)^m$ put

$$(2.2.21) \qquad g(x) = Q_x\{\Sigma G(y_{t_1-}, y_{t_1}, \ldots, y_{t_m-}, y_{t_m})\},$$

$$(2.2.22) \qquad f(y) = Q_y^*\{\Sigma\, F(y_{s_n}^*, y_{s_n-}^*, \ldots, y_{s_1}^*, y_{s_1-}^*)\}$$

where the sum in (2.2.21) is taken over all $t_1 < t_2 < \cdots < t_m$, $t_1, t_2, \ldots, t_m \in J$; and the sum in (2.2.22) is taken over all $s_1 < s_2 < \cdots < s_n$, $s_1, s_2, \ldots, s_n \in J$. Let M be a t.i. $(0,\Pi)$-generated set and let $\Sigma^{(k)}$ denote the sum over all k-tuples $\gamma_1, \gamma_2, \ldots, \gamma_k$ such that $\gamma_1 < \gamma_2 < \cdots < \gamma_k$. Then

$$Q\{\Sigma^{(m+n)} F(\gamma_1, \delta_1, \gamma_2, \delta_2, \ldots, \gamma_n, \delta_n) G(\gamma_{n+1}, \ldots, \gamma_{n+m}, \delta_{n+m})\}$$

$$= Q\{\Sigma^{(n)} F(\gamma_1, \delta_1, \gamma_2, \delta_2, \ldots, \gamma_n, \delta_n) g(\delta_n)\}$$

$$= Q\{\Sigma^{(m)} f(\gamma_1) G(\gamma_1, \delta_1, \ldots, \gamma_m, \delta_m)\}.$$

The proof of this lemma follows from the Theorem 2.1.1 and the strong Markov property of y_t and y_t^*.

3. Stationary Markov Processes with Infinite Underlying Distributions and their Subprocesses.

3.1. In this paper we deal with (general) Markov processes with random birth and death times and it is worthwhile to give a precise definition of such processes. Let $(\Omega, \mathscr{F})$ be a measurable space and P be a

σ-finite measure on $\mathscr{F}$. Suppose that two measurable functions $\alpha(\omega)$ and $\beta(\omega)$ ($\alpha(\omega) < \beta(\omega)$) are given; and suppose that for each $t \in T$, $x_t(\omega)$ is a measurable mapping of the set $\{\alpha(\omega) < t < \beta(\omega)\}$ into a Borel space E. We say that (x_t, P) is a (homogeneous) Markov process if the measure $\nu_t(\Gamma) = P\{x_t \in \Gamma\}$ is σ-finite and there exists a transition function p such that

$$P\{x_{t_1} \in dx_1, x_{t_2} \in dx_2, \ldots, x_{t_n} \in dx_n, \alpha < t_1, \beta > t_n\}$$

$$= \nu_{t_1}(dx_1)p(t_2 - t_1, x_1; dx_2)\ldots p(t_n - t_{n-1}, x_{n-1}; dx_n).$$

If ν_t does not depend on t then the process (x_t, P) is stationary.

Consider a stationary Markov process $(x_t, \overline{P})$, and assume that x_t is conservative, i.e., $\overline{P}\{\alpha \neq -\infty\} = \overline{P}\{\beta \neq +\infty\} = 0$. Suppose that the state space E of this process is divided into two sets D and V and let $M' = \{t : x_t \in V\}$ and

$$(3.1.1) \qquad M = \overline{M'}$$

where the bar over a set means its closure. We denote by $]\gamma, \delta[$ an element of the set of all open intervals contiguous to M. For each path $x_.$ and each $]\gamma, \delta[$ we associate a trajectory w_δ^γ in D by the formula $w_\delta^\gamma(s) = x_s$, $\gamma < s < \delta$. The set of all trajectories in D with random birth time α and death time β is denoted by W. Suppose M satisfies 3.1.α below:

3.1.α. For each $s < t$ the set $M \cap]s,t[$ is $\mathscr{B}]s,t[\times \mathscr{F}_{]s,t[}$-measurable, where $\mathscr{F}_{]s,t[}$ is the completion with respect to the measure $\overline{P}$ of $\sigma(x_u, s < u < t)$, and $\mathscr{B}(I)$ is the Borel σ-field of the interval I.

Then it is possible to define a measure P on W in the following way (W is endowed with the Kolmogorov σ-field $\mathscr{G}$).

$$P\{A\} = \overline{P}\{\sum_{\gamma} 1_A(w_\delta^\gamma)\}. \tag{3.12}$$

The process $(w(s),P)$ is called a subprocess in D of the process $(x_t,\overline{P})$. Let ν, $\overline{p}$, $\overline{P}_x$ be respectively the one-dimensional distribution, the transition function and the transition probabilities of the process $(x_t,\overline{P})$.

Taking

$$A = \{w(t_1) \in \Gamma_1, w(t_2) \in \Gamma_2, \ldots, w(t_n) \in \Gamma_n\}, \Gamma_1, \Gamma_2, \ldots, \Gamma_n \subset D,$$

we get

$$\begin{aligned} &P\{w(t_1) \in \Gamma_1, \ldots, w(t_n) \in \Gamma_n\} \\ &= \overline{P}\{x_{t_1} \in \Gamma_1, \ldots, x_{t_n} \in \Gamma_n; s \notin M \text{ for all } t_1 < s < t_n\} \end{aligned} \tag{3.1.3}$$

whence

$$P\{w(t_1) \in \Gamma_1\} = \nu(\Gamma_1), \tag{3.14}$$

and moreover, P is a Markov measure with transition function p given by 1.1.γ. If the measure ν is σ-finite, then so is P, and if for each t

(3.1.5) $$\overline{P}\{x_t \in V\} = 0,$$

then the one-dimensional distribution of P is equal to that of $\overline{P}$ (namely to ν). In the sequel we shall consider only processes $(x_t, \overline{P})$ subject to (3.1.5). Put $\tau_s = \inf\{t > s : x_t \in V\}$; $\tau = \tau_0$. (Measurability of τ_s follows from 3.1.α). If

3.1.A. For each $x \in D$

$$\overline{P}_x\{\tau > t\} \to 0 \quad \text{as } t \to \infty,$$

then for each $x \in D$

(3.1.6) $$p(t,x;\, D) \to 0 \quad \text{as } t \to \infty.$$

If

3.1.B. For any set $\Gamma \subset D$ such that $\nu(t) < \infty$

$$\overline{P}\{x_s \in \Gamma,\ \tau > s) \to 0 \quad \text{as } s \to \infty,$$

then for any set Γ such that $P\{w(0) \in \Gamma\} < \infty$

(3.1.7) $$P\{w(s) \in \Gamma,\ \alpha < 0,\ \beta > s\} \to 0 \quad \text{as } s \to \infty.$$

If

3.1.C. For some $s > 0$

$$\overline{P}\{\tau < s\} < \infty$$

then

(3.1.8) $$P\{\alpha \le 0,\ 0 < \beta < s\} < \infty.$$

Let T_t be the semigroup generated by the transition function p. Note that (3.1.6) is true iff T_t is a SD-semigroup. The condition (3.1.7) holds iff ν is null-excessive measure; (3.1.8) is true iff ν is quasi-finite excessive with respect to T_t measure. If both (3.1.7) and (3.1.8) are satisfied then we say that the process (w(s),P) has a null-quasi-finite one-dimensional distribution.

Let Ω be the sample space of the process $(x_t,\overline{P})$ and $\mathscr{F}$ be the basic σ-field in Ω on which the measure $\overline{P}$ is defined, and which is supposed to contain all sets of $\overline{P}$-measure zero. Denote by $\mathscr{F}_s$ the completion with respect to $\overline{P}$ of $\sigma(x_u, u < s)$ and by $\mathscr{C}_s$ the completion with respect to $\overline{P}$ of the σ-field generated by the sets

$$(\tau_u < r\},\quad u,r < s.$$

(if the process x_t is regular, then $\mathscr{C}_s \subset \mathscr{F}_s$.) We say that the set D is regular for $(\overline{x}_t,P)$ if for $t > s$, $\mathscr{C}_s \vee \mathscr{F}_s$ and x_t are conditionally independent given x_s. (This definition certainly asumes $\mathscr{C}_s \subset \mathscr{F}$).

A Markov process (x_t^1,Q_1) with the state space $E_1 = D \cup V_1$ and a Markov process (x_t^2,Q_2) with the state space $E_2 = D \quad V_2$ are said to be equivalent, if the one-dimensional distributions of both processes are concentrated on D and their finite-dimensional distributions coincide.

The following theorems are similar to Theorems 1 and 2 in [9].

THEOREM 3.1.1 *Let* $(w(s),P)$ *be a stationary Markov process in the state space* D *with the transition function* $p(t,x;\,\cdot)$ *measurable in* (t,x) *and subject to* (3.1.6). *If the one-dimensional distribution of* P *is null-quasi-finite, then this process is a subprocess of a conservative stationary Markov process* $(x_t,\overline{P})$ *satisfying* 3.1.A - 3.1.C *for which* D *is a regular set.*

The set of all stationary Markov measures with transition function p is denoted by S(p).

THEOREM 3.1.2 *If* $(w(s),P)$ *satisfies the conditions of Theorem* 3.1.1 *and if in addition* P *is a minimal element of* S(p), *then the process* $(x_t,\overline{P})$ *is unique up to equivalence.*

3.2. In this section we prove Theorem 3.1.1. Consider the one-dimensional distribution ν of $(w(s),\overline{P})$. It was proved in [3] that any null-excessive ν can be represented in the form

$$(3.2.1) \qquad \nu = \int_0^{\infty} \nu^s ds,$$

where ν^s is an entrance law for p. we denote by P* a Markov measure on $\mathcal{G}$ with the transition function p and the one-dimensional distributions ν^s. Put

$$(3.2.2) \qquad \Pi(\Gamma) = P^*\{\beta \in \Gamma\}.$$

We want to use Π for the construction of a t.i. $(0,\Pi)$-generated set. To do that, we must be sure that Π, defined by (3.2.2), is a σ-finite measure subject to (2.1.1). The next three lemmas are aimed to prove this fact.

LEMMA 3.2.1 <u>For any</u> $u > 0$ <u>the measure</u> $\nu - \nu T_u$ <u>is finite</u>.

PROOF By our assumptions $\mu = \nu - \nu T_s$ is a finite measure for some $s > 0$. For each $r > 0$, $\mu T_r < \infty$, and for $u = ks$ we have

$$(3.2.3) \qquad \begin{aligned} \nu-\nu T_{ks} &= \nu-\nu T_s+\nu T_s-\nu T_{2s}+\cdots+\nu T_{(k-1)s}-\nu T_{ks} \\ &= \mu+\mu T_s+\cdots+\mu T_{(k-1)s}. \end{aligned}$$

Each summand in the right side of (3.2.3) is a finite measure; and so is $\nu - \nu T_{ks}$.

By virtue of (3.2.1)

$$(3.2.4) \qquad \nu - \nu T_u = \int_0^u \nu^t dt.$$

Hence if $u < ks$, then $\nu - \nu T_u < \nu - \nu T_{ks}$, and the lemma is proved.

Lemma 3.2.1 shows that $\nu^s(D)$ is finite for m-almost all $s > 0$ (m is the Lebesque measure). On the other hand for $t > s$

$$\nu^t(D) = \nu^t(1) = \nu^s T_{t-s} 1 < \nu^s(1) = \nu^s(D).$$

Therefore, $\nu^s(D)$ is finite for all $s > 0$ and is a decreasing function of s. Consequently

$$P^*\{\beta > s\} = P^*\{w(s) \in D\} = \nu^s(D) < \infty, \ s > 0. \tag{3.2.5}$$

Formula (3.2.5) shows that the restriction on any interval $]s,\infty]$ of the measure Π, defined by (3.2.2), is a finite measure; as a result, Π is σ-finite.

LEMMA 3.2.2 *The measure* Π *defined by* (3.2.2) *satisfies* (2.1.1).

PROOF Put $f(s) = \nu^s(D)$.

$$\int_0^\infty x \wedge 1\Pi(dx) = \int_0^1 x\Pi(dx) + f(1) \tag{3.2.6}$$

$$= \iint_C dx\Pi(dy) + f(1),$$

where $C = \{(x,y) : x > 0, y > 0, x + y < 1\}$. By Fubini's Theorem (3.2.6) equals

$$\int_0^1 \{\int_y^1 \Pi(dy)\}dx + f(1) = \int_0^1 (\Pi R^x) - \Pi(R^1))dx + f(1)$$

$$= \int_0^1 (f(x) - f(1))dx + f(1)$$

$$= \int_0^1 f(x)dx = \int_0^1 \nu^x(D)dx.$$

By virtue of (3.2.4) the right side of the above formula is equal to $(\nu - \nu T_1)(D)$. Lemma 3.2.1 implies that this expression is finite.

By virtue of Theorem 2.1.1 and Lemma 3.2.2 we are able to construct a $(0,\Pi)$-generated translation invariant set M, subject to (2.1.1). Let $\tilde{\Omega}$ be the corresponding sample space, $\mathcal{F}$ be the corresponding σ-field on $\tilde{\Omega}$ and Q be the corresponding measure.

LEMMA 3.2.3 <u>For any function</u> f <u>on</u> T × T

$$P\{f(\alpha,\beta)\} = Q\{\sum_{\gamma} f(\gamma,\delta)\}. \tag{3.2.7}$$

PROOF Denote by P_t^* the t-shift of measure P^*, that is

$$P_t^*\{w(s_1) \in \Gamma_1,\ldots,w(s_n) \in \Gamma_n\} \tag{3.2.8}$$

$$= P^*\{w(s_1 - t) \in \Gamma_1,\ldots,w(s_n - t) \in \Gamma_n\}.$$

By virtue of (3.2.1), for every G-measurable function f

$$P\{f\} = \int_{-\infty}^{\infty} P_t^*\{f\}dt.$$

Hence,

$$P\{f(\alpha, \beta)\} = \int_{-\infty}^{\infty} P_t^* \{f(\alpha,\beta)\}dt \tag{3.2.9}$$

$$= \int_{-\infty}^{\infty} P_t^*\{f(t,\beta)\}dt$$

$$= \int_{-\infty}^{\infty} P^*\{f(t,\beta + t)\}dt$$

$$= \int_{-\infty}^{\infty} \{\int_{0}^{\infty} f(t,y + t)\Pi(dy)\}dt.$$

By virtue of (2.1.4), the right side of (3.2.9) is equal to the right side of (3.2.7).

Consider a measure $\overline{N}$ on $T \times T \times W$ defined below

$$\overline{N}(\Gamma \times \Delta \times A) = P\{\alpha \in \Gamma,\ \beta \in \Delta,\ w \in A\},\ \Gamma,\ \Delta \subset T,\ A \in G.$$

Put

(3.2.10) $$N(B) = \overline{N}(B \times W),\quad B \subset T \times T.$$

LEMMA 3.2.4 The measure N, defined by (3.2.10) is σ-finite.

PROOF If Π satisfies (2.1.1), then for $t > 0$

$$\Pi(R^t) < \infty.$$

The support of measure N is the set $C = \{(x,y) : y < x\}$. The set C may be represented as a countable union of rectangles $R =]u,v[\times]r,q[$, where $u < v < r < q$. For such rectangle R

$$N(R) = P\{u < \alpha < v,\ r < \beta < q\}$$

$$= \int_{u}^{v} P_t^*\{r < \beta < q\}dt$$

$$< \int_u^v P_t^*\{\beta > r\}dt$$

$$< \int_u^v P^*\{\beta > r - v\}dt$$

$$= (v - u)\Pi(R^{r-v})$$

$$< \infty.$$

Let $(X, \mathscr{A})$ and $(Y, \mathscr{B})$ be two measurable spaces and Q be a measure on $\mathscr{A}$. We say that $n(x; \Gamma)$, $x \in X$, $\Gamma \in \mathscr{B}$ is a stochastic Q-quasi kernel from X into Y if the following conditions are satisfied:

3.2.α. for any $\Gamma \in \mathscr{B}$ $n(\cdot; \Gamma)$ is $\mathscr{A}$-measurable;

3.2.β. for Q-almost all $x \in X$ $n(x; Y) = 1$;

3.2.γ. If Γ_i is a sequence of disjoint sets then

$$n(x; \bigcup_k \Gamma_k) = \sum_k n(x; \Gamma_k)$$

for Q-almost all $x \in X$.

Note that if $\overline{Q}$ is any measure on the product $X \times Y$ and a σ-finite measure Q on X is a projection of $\overline{Q}$ on X then the function $n(x; A)$ which is a Radon-Nikodym derivative of $\overline{Q}(dx \times A)$ with respect to $Q(dx)$ is a stochastic Q-quasi kernel from X into Y.

LEMMA 3.2.5 *Let* $(X, \mathscr{A})$ *and* $(Y, \mathscr{B})$ *be two measurable spaces and* Q *be a* σ-*finite measure on* X. *If* $n_1, n_2, \ldots$ *is a sequence of stochastic* Q-*quasi kernels from* X *into* Y, *then there exists a measure* $\overline{Q}$ *on* $(X \times Y^\infty, \mathscr{A} \times \mathscr{B}^\infty)$ *such that for any* k,

$$Q(\Delta \times \Gamma_1 \times \ldots \times \Gamma_k \times Y \times Y \times \ldots) = \int_\Delta n_1(x; \Gamma_1)$$

$$\ldots n_k(x; \Gamma_k) Q(dx), \qquad \Delta \in \mathscr{A}, \Gamma_i \in \mathscr{B}.$$

The proof of this lemma does not differ from the proof of the Kolmogorov theorem.

Fix $A \in G$ and let

$$m(x,y; A) = \frac{d\overline{N}(dx \times dy \times A)}{dN(dx \times dy)}.$$

Let $r_1, r_2, \ldots, r_k$ be a sequence of all rational numbers and

$$x(k) = x(k,\tilde{\omega}) = \sup \{t < r_k : t \in M\}$$

$$y(k) = y(k,\tilde{\omega}) = \inf \{t > r_k : t \in M\}$$

$$z(k) = z(k,\tilde{\omega}) = (x(k), y(k)).$$

LEMMA 3.2.6 *For any* k, $n_k(\tilde{\omega}; A) = m(z(k); A)$ *is a stochastic* Q-*quasi kernel from* $\tilde{\Omega}$ *into* W.

For the proof see Lemma 3.3.2 of [9].

Put

$$\Omega = \tilde{\Omega} \times W^{\infty}, \qquad \mathscr{F} = \tilde{\mathscr{F}} \times G^{\infty}.$$

By Lemma 3.2.5 there exists a measure $\overline{P}$ on Ω such that

$$\overline{P}\{A \times B_1 \times B_2 \times \cdots \times B_k \times W \times W \ldots\}$$

$$= \int_A n_1(\tilde{\omega};\, B_1) n_2(\tilde{\omega};\, B_2) \cdots n_k(\omega;\, B_k) Q(d\tilde{\omega}).$$

Let $E = D \cup V$, where V is a singleton. Let

$$k(t) = k(t,\omega) = \min\, \{k : x(k) < t < y(k)\}. \tag{3.2.11}$$

Define

$$x_t(\omega) = x_t(\tilde{\omega}, w_1, w_2, \ldots,) = \begin{cases} V & \text{if } t \in M(\tilde{\omega}) \\ w_{k(t)}(t) & \text{otherwise} \end{cases}$$

We want to establish that $(x_t, \overline{P})$ is a conservative stationary Markov process which is covering for $(w(s), \overline{P})$, and for which D is a regular set.

3.3. The following lemma establishes an important relation between measure Q, and quasi kernel m and measure P.

LEMMA 3.3.1 <u>For any functions</u> f <u>and</u> g <u>on</u> T <u>and any</u> $A \in G$

$$Q\{\sum_{\gamma} f(\gamma) g(\delta) m(\gamma, \delta;\, A)\} = P\{f(\alpha) g(\beta) 1_A\}. \tag{3.3.1}$$

PROOF We can apply Lemma 3.2.3 to the left side of (3.3.1) and obtain

$$\sum_{\gamma} f(\gamma)g(\delta)m(\gamma,\delta;\, A) \tag{3.3.2}$$

$$= \int_{T\times T} f(x)g(y)m(x,y;\, A)P\{(\alpha,\beta) \in (dx,dy)\}$$

$$\int_{T\times T} f(x)g(y)m(x,y;\, A)N(dx,dy),$$

where N is the measure defined by (3.2.10). Since $m(x,y;\, A)$ is the Radon-Nikodym derivative of $\overline{N}$ with respect to N, the right side of (3.3.2) may be rewritten as

$$\int_{T\times T\times W} f(x)g(y)1_A(w)\overline{N}(dx,dy,dw) = P\{f(\alpha)g(\beta)1_A\},$$

and that is equal to right side of (3.3.1).

COROLLARY <u>The process</u> $(w_t,\overline{P})$ <u>is a covering for</u> $(w(s),P)$.

PROOF By the construction, $x(k,\tilde{\omega}) \neq x(m,\tilde{\omega})$ if $k \neq m$, and

$$\bigcup_k\,]x(k),y(k)[\ = \bigcup_{\gamma}]\gamma,\delta[$$

where $]\gamma,\delta[$ are the intervals contiguous to M. Therefore, for each $\gamma(\tilde{\omega})$ there exists $m(\tilde{\omega})$ such that $w_\delta^\gamma(t) = w_{m(\tilde{\omega})}(t)$ for all $\gamma < t < \delta$. (To find such m we can take any $t \in\ ]\gamma,\delta[$ and put $m(\tilde{\omega}) = k(t,\tilde{\omega})$, where $k(t,\tilde{\omega})$ is given by

(3.2.11)). Therefore,

$$(3.3.3)\qquad \overline{P}\{\sum_{\gamma} 1_A(w_\delta^\gamma)\} = \overline{P}\{\overline{P}\{\sum_{\gamma} 1_A(w_\delta^\gamma)\,|\,\tilde{\mathscr{F}}\}\}$$

$$= Q\{\sum_{k} m(z(k);\ A)\}$$

$$= Q\{\sum_{k} m(\gamma,\delta;\ A)\} = P\{A\}.$$

The last equality in (3.3.3) is due to (3.3.1).

LEMMA 3.3.3 _The process_ $(x_t,\overline{P})$ _is a conservative stationary Markov process. The subprocess in_ D _of_ $(x_t\overline{P})$ _is equal to_ $(w(s),P)$.

PROOF For each x and y, $m(x,y;-)$ is a probability measure on W such that

$$(3.3.4)\qquad \alpha = x \text{ and } \beta = y \text{ a.s. } m(x,y,-).$$

Let

$$B(x,y) = \{w(\cdot) \in W : w(t) \text{ is undefined for some } t \in]x,y[\}.$$

(That is, if $w \in B(x,y)$, then the inequality $\alpha(w) < x < y < \beta(w)$ is not satisfied).

Let

$$A(a,b) = \{\omega : x_t(\omega) \text{ is undefined for some } a < t < b\}.$$

Since for each $t \in M$, $x_t = V$,

$$\overline{P}\{A(-\infty,+\infty)\} = \overline{P}\{\bigcup_k A(x(k),y(k))\}$$

$$(3.3.5) \qquad = Q\{\sum_k m(x(k),y(k);\ B(x(k),y(k)))\}$$

$$= Q\{\sum_\gamma m(\gamma,\delta;\ B(\gamma,\delta))\}$$

By virtue of (3.3.4) $m(x,y;\ B(x,y)) = 0$, hence (3.3.5) equals zero. That shows that $(x_t,\overline{P})$ is a conservative process.

Markov property of $(x_t,\overline{P})$ was proved in Lemma 4.1.2 of [9], stationarity was shown in Section 4.2 of [9]. The last statement of Lemma 3.3.3 follows from the corollary to Lemma 3.3.2.

LEMMA 3.3.4 <u>The set</u> D <u>is a regular set for</u> $(x_t,\overline{P})$.

PROOF Let $u_1,u_2,\dots,u_k$, $v_1,v_2,\dots v_k$, $s_1,s_2,\dots,s_n < s < t$. We need to show that for each $\Gamma,\Gamma_1,\dots,\Gamma_n$, $\Delta \subset E$ there exists a function g on E such that

$$\overline{P}\{x_s \in \Gamma, x_{s_1} \in \Gamma_1,\dots,x_{s_n} \in \Gamma_n,$$

$$(3.3.6) \qquad x_t \in \Delta, \tau_{u_1} < v_1,\dots,\tau_{u_n} < v_n\}$$

$$= \overline{P}\{g(x_s); x_s \in \Gamma, x_{s_1} \in \Gamma_1,\dots,x_{s_n} \in \Gamma_n$$

$$\tau_{u_1} < v_1, \ldots, \tau_{u_n} < v_n\}.$$

For simplicity of calculations we consider only the case of $n = k = 1$, $u < v < s_1$. Since the one-dimensional distributions of $\overline{P}$ are concentrated on D we may consider only the case in which Γ, Γ_1 and Δ are subsets of D. Put

$$D(s,t) = \{w \in W : \alpha(w) < s < t < \beta(w)\},$$

$$E(s,t) = \{w \in W : \alpha(w) < s < \beta(w) < t\},$$

$$A = \{w \in W : w(s_1) \in \Gamma_1\},\ B = \{w \in W : w(s) \in \Gamma\},$$

$$C = \{w \in W : w(t) \in \Delta\}.$$

Denote by $\lambda_i(s,t)$ the indicator of the set $\{\gamma_i < s < t < \delta_i\}$, by $\delta_i(s,t)$ the indicator of the set $\{\gamma_i < s < \delta_i < t\}$, and by w_i the cut off $w_{\delta_i}^{\gamma_i}$, $i = 1,2,\ldots$.

$$\overline{P}\{x_{s_1} \in \Gamma_1, x_s \in \Gamma, x_t \in \Delta, \tau_u < v\}$$

$$= \overline{P}\{ \sum_{\gamma_1<\gamma_2} \delta_1(u,v)\lambda_2(s_1,t)1_{ABC}(w_2)$$

$$+ \sum_{\gamma_1<\gamma_2<\gamma_3} \delta_1(u,v)\lambda_2(s_1,s_1)1_A(w_2)\lambda_3(s,t)1_{BC}(w_3) \tag{3.3.7}$$

$$+ \sum_{\gamma_1<\gamma_2<\gamma_3} \delta_1(u,v)\lambda_2(s_1,s)1_{AB}(w_2)\lambda_3(t,t)1_C(w_3)$$

$$+ \sum_{\gamma_1<\gamma_2<\gamma_3<\gamma_4} \delta_1(u,v)\lambda_2(s_1,s_1)1_A(w_2)\lambda_3(s,s)1_B(w_3)\lambda_4(t,t)1_C(w_4)\}.$$

The first term in the right hand side of (3.3.7) is equal to

$$Q\{ \sum_{\gamma_1<\gamma_2} m(\gamma_1,\delta_1: E(u,v))m(\gamma_2,\delta_2; D(s_1,t)AB)\}$$

$$(3.3.8) \qquad = Q\{\sum_{\gamma} m(\gamma,\delta;D(s_1 t)ABC\phi(\gamma))\}$$

$$= P\{\phi(\alpha);\ ABCD(s_1,t)\}$$

$$= P\{\phi(\alpha)p(t - s;\ w(s),\Delta);\ AB,\alpha < s_1 < s < \beta\},$$

where

$$\phi(x) = Q_x^*\{\sum_{t\in J} m(y_t,y_{t-};\ E(u,v))\}.$$

(The first equality in (3.3.8) is due to Lemma 2.2.6, the second to Lemma 3.3.1, the last equality is due to the Markov property of (w(s),P).) Similarly, we get that the sum of the second, the third and the fourth term in (3.3.7) equals

$$P\{\phi(\alpha)\psi(w(s));\ AB,\alpha < s_1 < s < \beta\}$$

$$(3.3.9) \qquad + P\{\phi_1(\alpha)p(t - s;\ w(s),\Delta);\ B,\alpha < s < \beta\}$$

$$+ P\{\phi_1(\alpha)\psi(w(s));\ B,\alpha < s < \beta\},$$

where

$$\phi_1(x) = Q_x^*\{\sum_{\substack{r,z\in J \\ r<z}} m(y_r,y_{r-},AD(s_1,s_1))$$

$$m(y_z,y_{z-};\ E(u,v))\},\quad x \in T$$

(3.3.10)

$$\psi(x) = \int_0^\infty P_x\{\beta \in dy\}Q_y\{\sum_{r\in J} m(y_{r-},y_r;$$

$$w(t - s) \in \Delta)\}, \quad x \in D.$$

Adding (3.3.9) to (3.3.8), we get (3.3.6) with

(3.3.11) $$g(x) = p(t - s, x; \Delta) + \phi(x).$$

LEMMA 3.3.5 <u>The transition function of</u> $(x_t, \overline{P})$ <u>is</u>

(3.3.12)
$$\overline{p}(u,x;\Delta) = p(u,x;\Delta) + \int_0^\infty P_x\{\beta \in dy\}Q_y\{\int_0^{\sigma_u} P^*_{y_t}\{w(u) \in \Delta\}du \quad \text{if } \Delta \subset D.$$

$$\overline{p}(u,x;\ V) = 0.$$

PROOF The Kolmogorov-Chapman equation for $\overline{p}$ was verified in Section 2 of [10], as well as the fact that $\overline{p}(u,x;\ D) = 1$ for all $u > 0$ and $x \in E$.

Putting $t - s = u$ in (3.3.11), one can see that for ν-a.a. x $\overline{p}(u,x;\ \Delta) = g(x)$, and for the proof of (3.3.12), it is enough to verify equality between $\phi(x)$, given by (3.3.10), and the second term in the right hand side of (3.3.12) for ν-almost all x. Put

$$\Theta(y) = Q_y\{\sum_{r \in J} m(y_{r-}, y_r;\ w(u) \in \Delta\}, \quad y \in T.$$

Applying successively the Markov property of $(w(s),P)$, Lemma 3.3.1 and Lemma 2.2.6, we get

(3.3.13) $$\int_\Gamma \phi(x)\nu(dx) = P\{1_\Gamma(w(s))\phi(w(s))\}$$

$$= P\{1_\Gamma(w(s)),\theta(\beta)\}$$

$$= Q\{\sum_\gamma m(\gamma,s;\ w(s) \in \Gamma)\theta(\delta)\}$$

$$= Q\{\sum_\gamma m(\gamma,\delta;\ w(u) \in \Delta)\theta'(\gamma)\},$$

where

$$\theta'(y) = Q^*_y\{\sum_{r\in J} m(y_r,y_{r-};\ w(s) \in \Gamma)\}.$$

In view of Lemma 3.3.1, (3.3.13) equals

$$(3.3.14) \qquad P\{\theta'(\alpha)1_\Delta(w(u))\} = P\{\theta'(\alpha)\xi(\alpha);\ \beta > u\},$$

where

$$\xi(y) = 1_{y<u}P^*_y\{w(u) \in \Delta\}/P^*_y\{\beta > u\}.$$

Applying Lemma 3.3.1, Lemma 2.2.6 and again Lemma 3.3.1, we get that (3.3.14) equals

$$Q\{\sum_{\gamma<u<\delta} \theta'(\gamma)\xi(\gamma)\}$$

$$(3.3.15) \qquad = Q\{\sum_{\gamma_1<\gamma_2} m(\gamma_1,\delta_1;w(s) \in \Gamma)\xi(\gamma_2)\lambda_2(u,u)\}$$

$$= Q\{\sum_\gamma m(\gamma,\delta;\ w(s) \in \Gamma)\xi'(\delta)\}$$

$$= P\{1_\Gamma(w(s))\xi'(\beta)\},$$

where

$$\xi'(y) = Q_y\{\sum_{\gamma<u<\delta} \xi(\gamma)\}.$$

By virtue of Lemma 2.2.1

$$\xi'(y) = Q_y\{\int_0^\infty 1_{y_t<u}\xi_{y_t}\Pi(y_t; R^u)dt\}$$

$$= Q_y\{\int_0^{\sigma_u}\xi_{y_t}P^*_{y_t}\{\beta > u\}dt\}$$

$$= Q_y\{\int_0^{\sigma_u}P^*_{y_t}\{w(u) \in \Delta\}dt\}.$$

Substituting the expression for $\xi'(y)$ in (3.3.15), we see that for any set Γ

(3.3.16) $$\int_\Gamma \phi(x)\nu(dx) = \int_\Gamma \phi_1(x)\nu(dx),$$

where $\phi_1(x)$ is the second term in the right hand side of (3.3.12). Formula (3.3.16) implies

$$\phi(x) = \phi_1(x) \quad \text{a.e.} \quad \nu.$$

3.3. The proof of Theorem 3.1.2 does not differ from the proof of Theorem 2 in [9]. Suppose $(x_t\overline{P})$ is a Markov process in $D \cup V$ for which D is a regular set and for which the subprocess in D equals to $(w(s),P)$. If P is extreme in the class of all stationary measures with transition function p, then for each t the measure P_t^* defined by (3.2.8) is extreme in the class of all Markov measures with transition function p (see [3]). Heuristically, this means that there is only one entrance point from V into D. This observation implies that the local time of x_t at V is uniquely determined by P*. The inverse of this local time y_t is $(0,\Pi)$-process, Π given by (3.2.2). Using usual "last exit decomposition", we obtain that the two dimensional distributions of $\overline{P}$ are uniquely determined by P_t^*, ν and the functionals of y_t, i.e., by the process $(w(s),P)$.

4. Enhancing of Semigroups

4.1. Now we consider the semigroup T_t and the measure ν described in Theorem 1.

If T_t is a S-semigroup then there exists a transition function $p(t,x;\ \Gamma)$ such that

$$T_t f(x) = p(t,x;\ f)$$

(see [2], Theorem 2.1). If T_t is a dying semigroup then

$$p(t,x;\ D) \to 0 \quad \text{as } t \to \infty.$$

By the theorem of Kuznecov (see [6]) there exists a stationary Markov process $(w(s),P)$ with random birth and death times whose one-dimensional distribution is equal to ν and transition function is equal to p. (To construct such a process $(w(s),P)$ we need also to specify an excessive function h, but in our case $(h(x) \equiv 1.)$ The conditions of Theorem 1 imply that the one-dimensional distribution of P is null-quasi-finite. By Theorem 3.1.1 there exists a covering process $(x_t,\overline{P})$ with the state space $E = D \cup V$.

Let $\overline{p}(t,x;\Gamma)$ be the transition function of $\overline{P}$. From Section 3 we know that

$$\overline{p}(t,x;D) = 1 \quad \text{for all } t \text{ and } x \in D. \tag{4.1.1}$$

Therefore, $\overline{p}(t,x;-)$, considered as a kernel from D into D, is a transition function. By Theorem 3.1.1 $(x_t,\overline{P})$ is a stationary conservative process with the one-dimensional distribution ν; consequently, ν is invariant with respect to $\overline{p}$. The semigroup $\tilde{T}_t$ generated by $\overline{p}$ is the semigroup we are looking for. The properties 1.2.A - 1.2.C are automatically satisfied by any semigroup generated by a transtion function. The property 1.1.A is a consequence of (4.1.1); and (1.2.1) follows from the fact that $(x_t,\overline{P})$ is a covering process for $(w(s),P)$, for which 1.1.γ holds. Lemma 3.3.5 gives us the explicit expression for $\tilde{T}_t f(x)$ in terms of "internal" characteristics of ν and T_t (we make trivial transformations in (3.3.12) to obtain the formula below).

$$\tilde{T}_t f(x) = T_t f(x) + \int_0^\infty Q_0\{\int_0^{\sigma(t-z)} \nu^{y_s}(f)ds\}\mu_x(dz). \tag{4.1.2}$$

where μ_x is a measure on $]0,\infty[$ such that $\mu_x]r,u[= T_r 1(x) - T_u 1(x)$, the family ν^s is an entrance law with respect to T_t for which (3.2.1) holds, and (y_s,Q_0) is an increasing process with independent increments with translation constant 0 and the Levi measure Π such that

$$\Pi]r,u] = \nu^r(D) - \nu^u(D),$$

and $\sigma(u)$ is the first hitting time of $]u,\infty[$ by y_s.

It is interesting to compare the formula (4.1.2) with the result of Getoor (see [4], Theorem (8.1)). He solves the inverse problem, namely, he finds an invariant distribution ν for the transition function $\overline{p}$ given by the formula analogous to (3.3.12). The expression for ν he obtains is similar to (3.2.1).

4.2. The proof of Theorem 2 does not differ from the proof of Theorem 2 in [10]. Suppose T_t and $\tilde{T}_t$ satisfy the condition of Theorem 2 and p and $\overline{p}$ are the transition functions which generate these semigroups. The main idea of the proof is to construct a conservative stationary Markov process $(x_t,\overline{P})$ in the state space $E \supset D$ such that the one-dimensional distributions of $\overline{P}$ are concentrated on D and equal ν the transition function of $(x_t,\overline{P})$ equals $\overline{p}$, the set D is regular for x_t and the subprocess in D of $(x_t,\overline{P})$ has transition function p. In this case Theorem 2 is a consequence of Theorem 3.1.2.

We start from a conservative stationary Markov process in the state space D with one-dimensional distribution ν and transition function $\overline{p}$. A multiplicative functional α_t is constructed in such a way that

$$p(s,x;\,\Gamma) = \overline{P}_x\{1_\Gamma(x_s)\alpha_s\}.$$

Let Ω be the sample space of the process x_t. We put $\tilde{\Omega} = \Omega \times (T)^\infty$ and construct a measure Q on $\tilde{\Omega}$ and a family of random variables $\tau_s(\tilde{\omega})$ in such a way that

4.2.A. The marginal distribution of Q on Ω is equal to $\overline{P}$.

4.2.B. For $t > s$ the conditional probability of τ_s to be greater than t, given ω is equal to $\alpha_{t-s}(\theta_s\omega)$.

4.2.C. For $t > s$ $\tau_s = \tau_t$ on the set $\{\tau_s > t\}$.

4.2.D. The σ-field $\mathscr{C}_s \vee \mathscr{F}_s$ and the pair (x_t, τ_t) are conditionally independent, given x_s, where $\mathscr{C}_s$ is the minimal σ-field in $\tilde{\Omega}$ generated by the sets $\{\tau_r < u;$ $r, u \leq s\}$.

The family τ_s has the same properties as the family of the first hitting times of a set in the state space. We put $M(\tilde{\omega})$ to be the closure of the set of values of the function $\tau_{\cdot}(\tilde{\omega})$. We put $x_t^*(\tilde{\omega})$ to be equal to $x_t(\omega)$ if $t \,\overline{\in}\, M(\tilde{\omega})$ and $x_t^*(\tilde{\omega}) \in V$ otherwise. In the same way as in [10] one can show that (x_t^*, Q) and $(x_t, \overline{P})$ has the same finite-dimensional distributions and the subprocess in D of (x_t^*, Q) is equal to $(w(s), P)$ where P is a Markov measure with the one-dimensional distribution ν and the transition function p. Now we need only to apply Theorem 3.1.2 to obtain the final result.

References

[1] Dellacherie, C. (1972), Capacités et Processus Stochastique, Berlin-Heidelberg - New York: Springer.

[2] Dynkin, E. B. (1965), Markov Processes, 1 and 2, Berlin-Heidelberg - New York: Springer.

[3] Dynkin, E. B. (1980), "Minimal Excessive Measures and Functions," Transactions of American Mathematical Society, 258, No. 1, 217-244.

[4] Getoor, R. K. (1979), "Excursions of a Markov Process," Annals of Probability, 7, No. 2, 244-266.

[5] Karlin, S. and S. M. Taylor (1981), A First Course in Stochastic Processes, Second Edition, Academic Press, New York.

[6] Kuznecov, S. E. (1973), "Construction of Markov Processes with Random Times of Birth and Death," Theory of Probability and its Applications, XVIII, 1, 571-575.

[7] Maisonneuve, B. (1983), "Ensembles Regeneratifs de la Droite," Z. fur W. Theorie und Ver. Geb.), 63, 501-510.

[8] Taksar, M. I. (1980), "Regenerative Sets on Real Line," Seminaire de Probabilitiés, XIV, Lecture Notes in Mathematics, 784, Springer.

[9] Taksar, M. I. (1981), "Subprocesses of Stationary Markov Processes," Z. fur W. Theorie und Ver. Geb., 55, 275-299.

[10] Taksar, M. I. (1983), "Enhancing of Semigroups," Z. fur W. Theorie und Ver. Geb., 63, 445-462.

M. I. Taksar
Department of Statistics
Florida State University
Tallahassee, Florida 32306

Seminar on Stochastic Processes, 1985
Birkhäuser, Boston, 1986

BROWNIAN OCCUPATION MEASURES ON COMPACT MANIFOLDS

by

Gunnar A. Brosamler

Let M be a compact C^∞ manifold of dimension d. A C^∞ metric $\mathbf{g}$ and a C^∞ vector field V on M determine a "Brownian motion" on M, i.e. a strong Markov process

$$X = \{\Omega, \mathscr{A}; P^x, x \in M; X_t : \Omega \to M, \mathscr{F}_t, \theta_t, t > 0\}$$

with continuous sample paths and generator $L = \frac{1}{2}\Delta_{\mathbf{g}} + V$, where $\Delta_{\mathbf{g}}$ stands for the Laplace-Beltrami operator, associated with $\mathbf{g}$. (If necessary we indicate by the subscript $\mathbf{g}$, that an object is associated with the metric $\mathbf{g}$.) Notice that the metric and the vector field can be recovered from the Brownian motion, from its generator L, to be precise.

It is well-known that the Brownian motion X on M is recurrent. This recurrence is closely connected with the existence of a unique X-invariant probability measure λ on M. The measure λ is absolutely continuous with respect to the Riemann measure $m_{\mathbf{g}}$ on M and a version ϕ of its density $d\lambda/dm_{\mathbf{g}}$ is in $C^\infty(M)$. Moreover, $\phi > 0$ and $L'\phi = 0$, where

$L' = \frac{1}{2}\Delta_g - V - \text{div}_g V$ is the m_g-dual of L. The measure λ permits the introduction of the λ-dual L^* of L, i.e. $L^* = \frac{1}{2}\Delta_g - V + \text{grad}_g(\log \phi)$, which is the generator of the dual Brownian motion X^*. The Brownian motion X is called symmetric if L is symmetric i.e. if $L = L^*$. In any case the symmetrized operator $L^s = \frac{1}{2}(L + L^*)$ is symmetric and thus generates a symmetric Brownian motion X^s. Notice that the X-invariant measure is also X^*- and X^s-invariant. According to [8], $L = L^* = L^s$ iff V is a gradient field (with respect to **g**), in which case $V = \frac{1}{2}\text{grad}_g(\log \phi)$. For general V we have $\text{div}_g(\phi(V - \frac{1}{2}\text{grad}_g(\log \phi))) = 0$, which is just another way of writing $L'\phi = 0$.

Since symmetric Brownian motions play an important role in the Corollary below, it should be pointed out that the result quoted from [8] implies immediately that for any smooth metric **g** and any smooth positive probability measure λ on M, the class of Brownian motions on M for which **g** is the metric and for which λ is the invariant probability measure, contains exactly one symmetric Brownian motion, namely the one with the generator $L = \frac{1}{2}\Delta_g + \frac{1}{2}\text{grad}_g(\log\frac{d\lambda}{dm_g})$.

The aspect of recurrence that we shall discuss in this paper, is the asymptotic behavior of the occupation measure L_t of a Brownian motion.

Recall that $L_t(\omega)$ is a random measure on M, defined as the image measure of Lebesgue measure on [0,t] under $X_\cdot(\omega)|_{[0,t]}$. It acts on measurable functions f on M in the usual way, i.e.

$$L_t(\omega)f = \int_M f(x)L_t(\omega,dx) = \int_0^t f(X_s(\omega))ds.$$

The last identity, whenever either integral is defined, is just a transformation of variables. The measure $L_t(\omega)$ is highly singular and its singularity increases with the dimension of M. Its support is the range of $X.(\omega)|_{[0,t]}$, which has Riemann measure 0 for $d > 2$. It would be interesting to know if for $d > 2$, $L_t(\omega)$ is the h-Hausdorff measure, restricted to the range of $X.(\omega)|_{[0,t]}$, for a suitable function h. This is certainly the case if g is flat. (See [3], [10]). In any case the latter example suggests that the Hausdorff dimension of the support of $L_t(\omega)$ is 2 for $d > 2$. Be that as it may, as $t \to \infty$, the normalized measures $t^{-1}L_t(\omega)$ converge to the smooth measure λ, in a sense to be made precise below.

In [1] we approached the singularity and the asymptotic behavior of $L_t(\omega)$ via the (real) Sobolev spaces $H^{\alpha}(M)$, $\alpha \in R$. For $\alpha > 0$, the Hilbertian spaces $H^{\alpha}(M)$ are subspaces of the Hilbertian space $L^2(M)$, and the Hilbertian space $H^{-\alpha}(M)$ is the dual of $H^{\alpha}(M)$. Moreover, $H^0(M)$ and $H^{-0}(M)$ are identified with the Hilbert space $L^2(M,\lambda)$. Notice that S_λ, defined by $S_\lambda f = \int_M f d\lambda$, is an element of $H^{-0}(M)$, or even of $H^{\infty}(M)$.

Strictly speaking the elements of $H^{\alpha}(M)$, $\alpha > 0$, are equivalence classes of functions rather than functions. Two functions in an equivalence class differ at most on a set of measure 0 (with respect to any smooth positive measure). As $L_t(\omega)$ is singular (with respect to any smooth positive measure) if $d > 2$, $L_t(\omega)$ is not an invariant of the equivalence class containing f if $d > 2$. In order to obtain such an invariant we shall apply L_t to a "regularized" version $\tilde{f}$ of f.

For a function $f \in L^1(M)$, the set $M_f \subseteq M$ and the function $\tilde{f} : M_f \to [-\infty,+\infty]$ are defined as follows. Let

$$M_f = \{x \in M;\ \lim_{r\to 0}[m_g(B_g(x,r))]^{-1} \int_{B_g(x,r)} f\, dm_g$$

exists and is the same for all smooth metrics g on $M\}$

and let $\tilde{f}$ be the limit in $\{\}$. Here B_g is the g-geodesic ball with center x and radius r. By a well-known theorem of Lebesgue, M_f^c is a null-set and $f = \tilde{f}$ a.e. (with respect to any smooth positive measure). Obviously, $M_{\tilde{f}} = M_f$, $\tilde{\tilde{f}} = \tilde{f}$, and if $f_1 = f_2$ a.e. then $M_{f_1} = M_{f_2}$, $\tilde{f}_1 = \tilde{f}_2$. Moreover, for $f \in C(M)$ we have $M_f = M$ and $\tilde{f} = f$.

For $f \in L^1(M)$, we define Ω_f as the set of all $\omega \in \Omega$ such that $X_t(\omega) \in M_f$ a.a. $t > 0$ and $\tilde{f} \circ X.(\omega) \in L^1_{loc}(R^+)$ or equivalently $\tilde{f} \in \bigcap_{t>0} L^1(M, L_t(\omega))$. Obviously, if $f_1 = f_2$ a.e. then $\Omega_{f_1} = \Omega_{f_2}$.

We now define for $t > 0$, $\omega \in \Omega$, $f \in H^\infty(M) = \varprojlim H^k(M)$ (i.e. $\tilde{f} \in C^\infty(M)$)

$$\tilde{L}_t(\omega)f = L_t(\omega)\tilde{f}$$

We observe that for $t > 0$, $\omega \in \Omega$, for any dimension d, $\tilde{L}_t(\omega) \in H^{-\infty}(M) = \varinjlim H^{-k}(M) =$ dual of $H^\infty(M)$, and more precisely that for $\alpha > d/2$, $\tilde{L}_t(\omega) \in H^{-\alpha}(M)$ and $\tilde{L}_t(\omega)f = L_t(\omega)\tilde{f}$ for $f \in H^\alpha(M)$. This follows immediately from the Sobolev embedding theorem, which states that $\tilde{H}^\alpha(M) \subseteq C(M)$ for $\alpha > d/2$, where the embedding is continuous (even compact). The situation is less trivial for indices $\alpha < d/2$ if $d > 2$. If $d = 1$, the existence of a continuous local time assures that $\tilde{L}_t$ is at least in $H^0(M)$. The case

$\alpha < d/2$, $d > 2$ has been studied in [1]. The key tool in that paper is the observation that the operators L, L^*, L^s are Hilbertian space isomorphisms $L, L^*, L^s : H^2_\lambda(M) \to L^2_\lambda(M)$, where $H^\alpha_\lambda(M) = \{f \in H^\alpha(M);\ S_\lambda f = 0\}, \alpha > 0$; $L^2_\lambda(M) = \{f \in L^2(M);\ S_\lambda f = 0\}$.

In the present paper we establish the following two theorems.

THEOREM 1: <u>There exists a shift-invariant set</u> $\Omega_0 \subseteq \Omega$ <u>of full</u> P^x<u>-measure, all</u> $x \in M$, <u>such that</u>

(1) <u>For all</u> $\omega \in \Omega_0$, <u>the following hold.</u>

(a) $\tilde{L}_t(\omega) \in H^{-\alpha}(M)$, <u>all</u> $t > 0$, <u>all</u> $\alpha > d/2 - 1$.

(b) <u>If</u> $d > 2$, <u>we have for all</u> $f \in \bigcup_{\alpha > d/2-1} H^\alpha(M)$ <u>that</u> $\omega \in \Omega_f$ <u>and</u> $\tilde{L}_t(\omega)f = L_t(\omega)\tilde{f}$, $t > 0$. <u>If</u> $d = 1$, <u>the same is true for</u> $f \in H^0(M)$, <u>and in this case</u> $L_t(\omega)\tilde{f} = L_t(\omega)f$, $t > 0$.

(c) $\tilde{L}_t(\omega) \notin H^{-(d/2\ -1)}(M)$, <u>all</u> $t > 0$.

(d) <u>If</u> $d > 2$, <u>there exists a function</u> $f \in H^{d/2\ -1}(M)$, $f > 0$ <u>such that</u> $M_f = M$, $\tilde{f} = f$ <u>and</u> $\int_{0^+} f \circ X_s(\omega)dx = \infty$.

(2) <u>For all</u> $\alpha > d/2 - 1$, <u>the process</u> $\{\tilde{L}_t,\ t > 0\}$ <u>on</u> Ω_0 <u>is a strongly continuous</u> $H^{-\alpha}(M)$<u>-valued additive functional, measurable with respect to the Borel sets in</u> $H^{-\alpha}(M)$.

THEOREM 2: <u>For all</u> $x \in M$, P^x - a.e. <u>for all</u> $\alpha > d/2 - 1$

(a) <u>The random set</u>

$$\left\{ \frac{\tilde{L}_t(\omega) - tS_\lambda}{\sqrt{2t \log \log t}},\quad t > e^2 \right\}$$

in $H^{-\alpha}(M)$ is relatively strongly compact in $H^{-\alpha}(M)$.

(b) The strong $H^{-\alpha}(M)$-cluster set as $t \to \infty$, of

$$\frac{\tilde{L}_t(\omega) - tS_\lambda}{\sqrt{2t \log \log t}}$$

equals

$$\{f \in H^1(M);\ \int_M f\, d\lambda = 0,\ \int_M |\mathrm{grad}_g f|^2 d\lambda$$

$$+ \int_M |\mathrm{grad}_g (L^s)^{-1} V_1 f|^2 d\lambda \le 4\},$$

where $V_1 = V - \frac{1}{2}\,\mathrm{grad}_g(\log \phi)$.

Theorem 2(b) implies immediately the following COROLLARY (MAXIMUM PRINCIPLE): Let g be a smooth metric and λ a smooth, positive probability measure on M. Considering the class of all Brownian motions on M with metric g and invariant probability measure λ, the cluster set as $t \to \infty$, of

$$\frac{\tilde{L}_t - tS_\lambda}{\sqrt{2t \log \log t}}$$

is maximal for the symmetric Brownian motion in this class. THe maximal cluster set equals $\{f \in H^1(M);$ $\int_M f\, d\lambda = 0,\ \int_M |\mathrm{grad}_g f|^2 d\lambda \le 4\}$.

For very general laws of the iterated logarithm (LIL) for Banach space - valued random variables see also [5], [6]. Here our Banach space is $H^{-\alpha}(M)$, and we are able to give a more explicit description of the cluster set in this special case using the differential structure of the manifold.

Parts of theorem 1 and 2 were proved in [1]. That paper improves results in [2]. It does not contain (1)(c) and (1)(d) of Theorem 1 and hence does not give the precise degree of singularity of $L_t(\omega)$. As well it gives as cluster set in Theorem 2 the set $(G + G^*)^{1/2} B$, where G and G^* are the Green operators of X and X^* respectively, and B is the unit ball in $L^2(M,\lambda)$. It is the purpose of this paper to prove (1)(c) and (1)(d) of Theorem 1 as well as

THEOREM 3:

$$(G + G^*)^{1/2} = \{f \in H^1(M); \int_M |\mathrm{grad}_g f|^2 d\lambda$$

$$+ \int_M |\mathrm{grad}_g (L^s)^{-1} V_1 f|^2 d\lambda < 4, \int_M f\, d\lambda = 0\}.$$

It is clear that Theorem 3 in connection with the earlier results of [1] gives Theorem 2 of the present paper.

Before we proceed to the proofs, we shall make a few remarks. Regarding Theorem 1, (1)(c), it can be shown that for $\alpha > \max(0, d/2 - 2)$, $f \in H^\alpha(M)$, $x \in M$, we have $P^x(\Omega_f) = 1$. However it is the simultaneous integrability of $\tilde{f} \circ X.$ on Ω_0 for every $f \in H^\alpha(M)$, we are concerned with in Theorem 1, and this requires a more careful analysis.

Regarding Theorem 2, it has been pointed out in [1] that a LIL in $H^{-\alpha}(M)$ implies a LIL simultaneously for all $\tilde{f}$, $f \in H^\alpha(M)$, and a LIL, using a suitable norm in $H^{-\alpha}(M)$, will be discussed below. Notice that the cluster set in Theorem 2 is a closed bounded set in $H^1(M)$, and therefore a compact set in $H^\alpha(M)$ for all $\alpha < 1$. Moreover, the last integral in the description of the cluster set is a continuous functional on $L^2(M)$, if V_1 is considered as a generalized derivative and $(L^s)^{-1}$ is replaced by $G^s + S_\lambda$ where G^s is the Green operator of X^s. Notice that for $f \in H^1(M)$, $S_\lambda(V_1 f) = 0$, that V_1 is a continuous linear operator $V_1 : L^2(M) \to H^{-1}(M)$ and that by (3.4) in [1], $G^s + S_\lambda$ is a Hilbertian space isomorphism $G^s + S_\lambda :$ $H^{-1}(M) \to H^1(M)$. The operator $(G^s + S_\lambda)V_1 : L^2(M) \to H^1(M)$, by the way, is an operator with kernel $-V_1(y)\cdot \mathrm{grad}_{\mathbf{g}} g^s(x,y)$ (with respect to λ), where g^s is the kernel of G^s (with respect to λ). This follows from $\mathrm{div}_{\mathbf{g}}(\phi V_1) = 0$ and implies $(G_s + S_\lambda)V_1 : L^2(M) \to H^1_\lambda(M)$. The functional under consideration vanishes if V is the gradient field (so in particular if $V = 0$), i.e. if L is symmetric. It equals $4c^2\int_M f^2 dx$ for Brownian motion on S^1 with constant drift c d/dx.

It may be of interest that one can reformulate Theorem 1(2) and Theorem 2 so as to assert continuity and a LIL in the linear space $H = \bigcap_{\alpha > d/2-1} H^{-\alpha}(M)$, endowed with the smallest topology for which all embeddings $i : H \to H^{-\alpha}(M)$, $\alpha > d/2 - 1$, are continuous. Obviously, a net x_β in H converges to $x \in H$ iff x_β converges to x in $H^{-\alpha}(M)$ for all $\alpha > d/2 - 1$, or only for $\alpha = d/2 - 1 + 1/n$, all n. This implies immediately that $\tilde{L}. : \mathbf{R}^+ \to H$ is continuous, and if K denotes the cluster set of Theorem 2(b), that the set

$$K(\omega) = \{\frac{\tilde{L}_t(\omega) - tS_\lambda}{\sqrt{2t \log \log t}},\ t > e^2\} \cup K,$$

being closed (even compact) in each $H^{-\alpha}(M)$, $\alpha > d/2 - 1$, is closed in H. Using Theorem 2 we see by a diagonal argument that every sequence t_n converging to ∞, has a subsequence t_{n_k} such that

$$\frac{\tilde{L}_{t_{n_k}} - t_{n_k} S_\lambda}{\sqrt{2t_{n_k} \log \log t_{n_k}}}$$

converges in H, and that the H-cluster set as $t \to \infty$, of

$$\frac{\tilde{L}_t(\omega) - tS_\lambda}{\sqrt{2t \log \log t}}$$

equals K. It follows that the H-closure of

$$\{\frac{\tilde{L}_t(\omega) - tS_\lambda}{\sqrt{2t \log \log t}},\ t > e^2\}$$

equals $K(\omega)$. Compactness in H of $K(\omega)$, i.e. relative compactness in H of

$$\{\frac{\tilde{L}_t(\omega) - tS_\lambda}{\sqrt{2t \log \log t}},\ t > e^2\}$$

now follows from compactness of K in $H^\alpha(M)$, $\alpha < 1$, hence in H, and from the continuity of $\tilde{L}.: \mathbf{R}^+ \to H$. For the argument notice that for $K \subseteq 0$, 0 open in H, the set

$$\{t > e^2;\ \frac{\tilde{L}_t(\omega) - tS_\lambda}{\sqrt{2t \log \log t}} \notin 0\} \subseteq \mathbf{R}^+$$

is bounded and closed, hence compact.

Regarding the Corollary, we point out that it implies in particular: Given any smooth metric **g** on M, then within the class of Brownian motions on M with generators

$L = \frac{1}{2}\Delta_g + V$, where $\operatorname{div}_g V = 0$, the Brownian motion with generator $L = \frac{1}{2}\Delta_g$ has asymptotically the highest degree of randomness, in the sense that for it the cluster set as $t \to \infty$, of the process

$$\frac{\tilde{L}_t - tS_{m_g}}{\sqrt{2t \log \log t}}$$

is maximal.

Finally notice that for dimension $d = 1$, the statements of Theorem 1, (1)(a) and (2), Theorem 2 are somewhat stronger than the results stated in [1] for $d = 1$, since we admit here negative α. However the ncessary arguments are essentially contained in [1] (see e.g. Lemma (7.13) in [1]).

The methods used are potential theoretic, just as in [1]. The central role is played by the α-potential kernels $\{g_\alpha, \ \alpha > 0\}$, where

$$g_\alpha(x,y) = \frac{1}{\Gamma(\alpha)} \int_0^\infty t^{\alpha-1}\{p(t,x,y) - 1\}dt,$$

and p is the transition density of X with respect to λ. These kernels, or rather the kernels g_α^s corresponding to X^s, enter on the one hand in the description of the spaces $H^\alpha(M)$ by Theorem (3.2) in [1], and on the other hand in the definition of the α-potentials of the occupation measures L_t, i.e. $\int_M g_\alpha^s(y,\cdot)L_t(\omega,dy) = \int_0^t g_\alpha^s(X_\sigma,\cdot)d\sigma$. It is the square integrability of these α-potentials that characterizes "good" paths. We shall give two illustrations of these aspects. Firstly, according to Theorem (3.2) in [1], for every $\alpha > 0$, the $H^\alpha(M)$-functions f are just those functions, which coincide a.e. with the potentials $G^s_{\alpha/2}\bar{f} + c$ of functions $\bar{f} \in L^2(M)$, where $(G^s_\alpha \bar{f})(x) = \int_M g^s_\alpha(x,y)\bar{f}(y)d\lambda(y)$. It is these potentials that were used in [1] as "regularized" versions of f. We shall argue that they are essentially the same as the $\tilde{f}$, defined above. Notice first that for $\bar{f} \in L^1(M)$, $\bar{f} > 0$ we have

(1) $$M_{G^s_{\alpha/2}\bar{f}} = M,$$

(2) $$G^s_{\alpha/2}\bar{f} = G^s_{\alpha/2}\bar{f} :$$

If the averages of the $G^s_{\alpha/2}\bar{f}$ are taken with respect to the metric $\mathbf{g}$ that is used for the definition of the operators G^s_α, the existence of the limit and its identification follow from the equation

$$\lim_{r\to 0} E^x(G^s_{\alpha/2}\bar{f})(X_{\tau_{x,r}}) = (G^s_{\alpha/2}\bar{f})(x), \quad x \in M,$$

where $\tau_{x,r}$ is the first exit time from $B_{\mathbf{g}}(x,r)$. - In order to prove that the averages around x of $G^s_{\alpha/2}\bar{f}$, taken with respect to an arbitrary metric, converge to $(G^s_{\alpha/2}\bar{f})(x)$, one may use the notion of α-thinness. This notion is associated with the differentiable manifold, as the singularities of the operators G_α generated by different metrics, are equivalent. One makes use of the fact that there is a set $E \subseteq M$, which is α-thin in x and for which $\lim_{y\to x, y\notin E}(G^s_{\alpha/2}\bar{f})(y) = (G^s_{\alpha/2}\bar{f})(x)$. Then one uses the fact that for any set $E \subseteq M$, which is α-thin in x, $m(B_{\mathbf{g}}(x,r) \cap E)/m(B_{\mathbf{g}}(x,r)) \to 0$ for any smooth positive measure and any smooth metric $\mathbf{g}$ on M. Obviously (1), (2) still hold if "$\overline{f} > 0$" is replaced by "$G^s_{\alpha/2}\overline{f}^+ < \infty$" or "$G^s_{\alpha/2}\overline{f}^- < \infty$." For general $\overline{f} \in L^1(M)$, let

$$A_{\overline{f}} = \{x \in M;\ G^s_{\alpha/2}\overline{f}^+ < \infty \text{ or } G^s_{\alpha/2}\overline{f}^- < \infty\};$$

the $m(A^c_{\overline{f}}) = 0$ and

$$(1') \qquad A_{\overline{f}} \subseteq M_{G^s_{\alpha/2}\overline{f}}$$

$$(2') \qquad \widetilde{G^s_{\alpha/2}\overline{f}} = G^s_{\alpha/2}\overline{f} \quad \text{on } A_{\overline{f}}.$$

If now $f \in H^\alpha(M)$, $\alpha > 0$, i.e. $f = G^s_{\alpha/2}\overline{f} + c$ a.e. $\overline{f} \in L^2(M)$, it follows that

$$(1'') \qquad A_{\overline{f}} \subseteq M_f$$

$$(2'') \qquad \tilde{f} = G^s_{\alpha/2}\overline{f} + c \quad \text{on } A_{\overline{f}}.$$

In particular we conclude that for any $\omega \in \Omega$, such that

$$\int_M d\lambda(y)\left[\int_0^t d\sigma \left|g^s_{\alpha/2}(X_\sigma, y)\right|\right]^2 < \infty$$

for $t > 0$, we have

$$G^s_{\alpha/2}|\overline{f}| \quad X.(\omega) \in L^1_{loc}(R^+),$$

therefore $X_t(\omega) \in A_{\overline{f}} \subseteq M_f$ a.a. $t > 0$ and

$$L_t(\omega)\tilde{f} = L_t(\omega)(G^s_{\alpha/2}\overline{f}) + ct.$$

As a second illustration we discuss an amusing corollary to Theorem 2 in the case $V = 0$. Assume $\alpha > \max(0, d/2 - 1)$. By Theorem 1, $\tilde{L}_t \in H^{-\alpha}(M)$, and we know from [1], that there is an admissible norm $\| \ \|_{H^{-\alpha}(M)}$ in $H^{-\alpha}(M)$, for which

$$\|\tilde{L}_t - tS_m\|^2_{H^{-\alpha}(M)} = \int_M dm(x)\left(\int_0^t g_{\alpha/2}(x, X_s)\right)^2$$

$$= \int_0^t ds_1 \int_0^t ds_2 g_\alpha(X_{s_1}, X_{s_2}).$$

By a transformation of variables

$$\|\tilde{L}_t - tS_m\|^2_{H^{-\alpha}(M)} = \int_M \int_M g_\alpha(x,y) L_t(\omega, dx) L_t(\omega, dy),$$

which is the α-energy of the random measure $L_t(\omega)$. By Theorem 2(b) with the cluster set $\sqrt{2}\, G^{1/2} B$, we have for $x \in M$, P^x - a.e. for $\alpha > d/2 - 1$

$$\overline{\lim}_{t\to\infty} \frac{\|\tilde{L}_t(\omega) - tS_m\|^2_{H^{-\alpha}(M)}}{2t \log\log t}$$

$$= \sup\{2\|G^{1/2} f\|^2_{H^{-\alpha}(M)}, \int_M f^2 dm \leq 1\}$$

$$= 2(2/\lambda_0)^{\alpha+1},$$

where λ_0 is the smallest eigenvalue of $-\Delta$. In other words, the eigenvalue λ_0 can be obtained by observing the α-energy of a typical path for large t. If $d = 2$ or 3, $\alpha = 1$ is strictly larger than $d/2 - 1$, so the classical energy may be used.

For the proofs of Theorem 1, (1)(c) and (1)(d) and of Theorem 3 we recall some notations and results from [1].

We denote by $r(x,y)$ the geodesic distance between x and y, associated with $\mathbf{g}$. We shall omit the subscript $\mathbf{g}$ from now on. We denote by $\{g_\alpha^s, \ \alpha > 0\}$ the α-potential kernels corresponding to L^s and let $k_\alpha^s = g_{\alpha/2}^s + 1$. The $\{g_\alpha^s, \ \alpha > 0\}$ form a semigroup with respect to λ. They are symmetric, bounded below and satisfy $\sup_{x \in M} \int |g_\alpha^s(x,y)| d\lambda(y) < \infty$. The same statements are obviously true for k_α^s. We shall use the operators K_α^s, $\alpha > 0$; G, G^*, G^s on $L^2(M)$ defined by $(K_\alpha^s f)(x) = \int_M k_\alpha^s(x,y) f(y) d\lambda(y)$, $(Gf)(x) = \int_M g(x,y) f(y) d\lambda(y)$, $(G^* f)(x) = \int_M g(y,x) f(y) d\lambda(y)$, $(G^s f)(x) = \int_M g^s(x,y) f(y) d\lambda(y)$, where $g = g_1$, $g^s = g_1^s$ are the 1-potential kernels corresponding to L, L^s. According to theorem (3.2) of [1], the K_α^s, $\alpha > 0$, are Hilbertian space isomorphisms $K_\alpha^s : L^2(M) \to H^\alpha(M)$. Recall that $G1 = 0$, $S_\lambda(Gf) = 0$ and that the operators L and G are inverses of each other in the following sense

$$LG = GL = -I + S_\lambda$$

Here I denotes the identity operator in $L^2(M)$ or $H^2(M)$. (The same relations hold for L^*, G^* and L^s, G^s). It follows that if $V = 0$, $G_\alpha = (-\tfrac{1}{2}\Delta)^{-\alpha}$ on $L_\lambda^2(M)$.

<u>Proof of Theorem 1, (1)(c) and (1)(d) for d > 2:</u>

Assume first that $d = 2$. In this case the arguments are essentially those for the nonexistence of a local time: Let $\Omega_0 = \{\omega;\ m(X.(\omega)(\mathbf{R}^+)) = 0\}$. Then $P^x(\Omega_0) = 1$, $x \in M$, by a Fubini type argument, as the Brownian paths do not hit fixed points. As for (1)(d), it is sufficient to prove that for $\omega \in \Omega_0$, there exists a continuous (extended real-valued) function $f \in L^2(M)$, $f > 0$, which assumes the value ∞ on the compact set $K_\omega = X.(\omega)([0,1])$. Obviously $m(K_\omega) = 0$. Now let $\{O_n, n > 1\}$ be a decreasing sequence of open sets in M, such that $\bigcap_{n>1} O_n = K_\omega$ and $m(O_n) < 1/4$. There exist functions $f_n \in C(M)$, $n > 1$, such that $0 < f_n < 1$, $f_n = 0$ on O_n^c, $f_n = 1$ on K_ω. The function $f = \Sigma_{n=1}^{\infty} f_n$ has the desired properties. - Assertion (1)(c) is proved as follows. If for $\omega \in \Omega_0$, there were a continuous extension of $L_t|_{C^\infty(M)}$ to $L^2(M)$, there would exist $u \in L^2(M)$ such that for $f \in C^\infty(M)$,

$$\int_M f(x)u(x)dm(x) = \int_M f(x)L_t(\omega,dx).$$

As the uniform closure of $C^\infty(M)$ is $C(M)$, it would follow that $L_t(\omega,d\cdot) = u(\cdot)dm(\cdot)$, $u > 0$. This however, is false, because if we let $K_\omega^t = X.(\omega)([0,t])$, then $L_t(\omega,K_\omega^t) = t$ and for $\omega \in \Omega_0$, $m(K_\omega^t) = 0$.

From now on let $d > 3$ and let $\alpha = d/2 - 1$. We define the following sets in Ω:

$$\Omega_0 = \{\omega; 1\{s > 0; \int_{s^+}[r(X_s,X_\sigma)]^{-2}d\sigma < \infty\} = 0\},$$

$$\Omega_1 = \{\omega; \exists\ f \in H^\alpha(M),\ f > 0, \text{ such that } M_f = M,\ \tilde{f} = f,\ \int_{0^+} f \circ X_s(\omega)ds = \infty\},$$

$$\Omega_t^1 = \{\omega; K_\alpha^s \bar{f} \circ X.(\omega) \in L^1([0,t]), \text{ all } \bar{f} \in L^2(M),\ \bar{f} > 0\},\quad t > 0,$$

$$\Omega_t^2 = \{\omega : L_t(\omega)\big|_{C^\infty(M)} \text{ has a continuous extension to } H^\alpha(M)\},\ t > 0$$

$$\Omega_t^3 = \{\omega;\ \int_0^t k_\alpha^s(X_\sigma,\cdot)d\sigma \in L^2(M)\},\ t > 0.$$

It suffices to prove

(1) $$P^x(\Omega_0) = 1, \text{ all } x \in M,$$

(2) (a) $\Omega_t^1 \subseteq \Omega_t^3$, $t > 0$, (b) $\Omega_t^2 \subseteq \Omega_t^3$, $t > 0$,

(3) $$\cup_{t>0}\Omega_t^3 \subseteq \Omega_0^c,$$

(4) $$\cap_{t>0}[(\Omega_t^1)^c] \subseteq \Omega_1,$$

as (4), (2)(a), (3) imply $\Omega_0 \subseteq \Omega_1$, and (3), (2)(b) imply $\Omega_0 \subseteq \cap_{t>0}[(\Omega_t^2)^c]$.

For the proof of (1) we observe first that by [7], p. 96, problem 1, we have for all $x \in M$, that P^x - a.e.

$$\overline{\lim}_{t\downarrow 0} \frac{r(X_0,X_t)}{\sqrt{2t \log|\log t|}} = 1$$

(Choose e.g. for the above reference a coordinate system around x, in which the metric in x is Euclidean.) This LIL implies that for all $x \in M$, P^x - a.e. $\int_{0^+}[r(X_0,X_s)]^{-2}ds = \infty$. We now obtain (1) by using the Markov property of the Brownian motion and Fubini's theorem.

As for (2)(a), we observe that for all $\overline{f} \in L^2(M)$, $\overline{f} > 0$

(*) $$L_t(\omega)(K_\alpha^s\overline{f}) = \int_0^t d\sigma(\int_M d\lambda(x)k_\alpha^s(X_\sigma,x)\overline{f}(x))$$

$$= \int_M d\lambda(x)\overline{f}(x)(\int_0^t d\sigma k_\alpha^s(X_\sigma, x)),$$

since k_α^s is bounded below, and therefore $\{\overline{f}(\cdot\cdot)k_\alpha^s(X.,\cdot\cdot)\}^-$ integrable on $[0,t] \times M$. By assumption we have for $\overline{f} \in L^2(M)$, $\overline{f} > 0$ that $K_\alpha^s\overline{f} \circ X.(\omega) \in L^1_{loc}([0,t])$. It follows that for all $\overline{f} \in L^2(M)$, $\overline{f} > 0$, hence for all $\overline{f} \in L^2(M)$, we have $\overline{f}(\cdot) \int_0^t d\sigma k_\alpha^s(X_\sigma,\cdot) \in L^1(M)$. (2a) now follows from a theorem of Lebesgue, by which for any measure space (M,m), $\overline{f}\cdot h \in L^1(M,m)$ for all $\overline{f} \in L^2(M,m)$ implies $h \in L^2(M,m)$. The latter theorem by the way, is an easy consequence of the Banach-Steinhaus Theorem.

In the case of (2)(b), (*) still holds for all $\overline{f} \in C^\infty(M)$, as

$$(**) \qquad \sup_{y \in M} \int |k_\alpha^s(y,x)| d\lambda(x) < \infty.$$

In this case there exists $u \in L^2(M)$ such that for all

$$\overline{f} \in C^\infty(M),\ L_t(\omega)(K_\alpha^s\overline{f}) = \int_M u(x)\overline{f}(x)d\lambda(x),$$

since K_α^s is a Hilbertian space isomorphism $K_\alpha^s : L^2(M) \to H^\alpha(M)$ and $K_\alpha^s(C^\infty(M)) = C^\infty(M)$. It follows that $u(x) = \int_0^t d\sigma k_\alpha^s(X_\sigma,x)$, in particular that $\int_0^t d\sigma k_\alpha^s(X_\sigma,\cdot) \in L^2(M)$.

For the proof of (3) observe that $\{k_\alpha^s(X.,\cdot\cdot\cdot)k_\alpha^s(X..,\cdot\cdot\cdot)\}^-$ is an integrable function on $[0,t] \times [0,t] \times M$, because of (**) and since k_α^s is bounded below. Applying Fubini's theorem and the semigroup property of $\{k_\alpha^s,\ \alpha > 0\}$ we obtain

$$0 < \int_0^t \int_0^t k^s_{2\alpha}(X_{\sigma_1}, X_{\sigma_2}) d\sigma_1 d\sigma_2 = \int_M d\lambda(x) \{\int_0^t k^s_\alpha(X_\sigma, x) d\sigma\}^2 < \infty.$$

(3) follows now from the following estimate: There exists $C > 0$ such that for $x, y \in M$

$$k^s_{2\alpha}(x,y) > C\{[r(x,y)]^{-d+2\alpha} - 1\} = C\{[r(x,y)]^{-2} - 1\},$$

which itself follows from the definition of $k^s_{2\alpha}$ in terms of the transition density and the following lower estimate for the transition density: There exist r_0, C, T such that

$$p(t,x,y) > Ct^{-d/2} \exp(-r(x,y)^2/2t)$$

for $r(x,y) < r_0$, $0 < t < T$. This latter estimate follows e.g. from Theorem 1 in [4].

Finally, to prove (4) let $\omega \in (\Omega^1_{1/n})^c$ for all $n \in \mathbf{N}$. There exist $\bar{f}_n > 0$, $\bar{f}_n \in L^2(M)$ such that

$$\|\bar{f}_n\|_{L^2(M,\lambda)} = 1, \int_0^{1/n} (K^s_\alpha \bar{f}_n)(X_\sigma(\omega)) d\sigma = \infty.$$

Let $\bar{f} = \Sigma 2^{-n} \bar{f}_n$, $f = K^s_\alpha \bar{f}$. Then $f \in H^\alpha(M)$. Since $\bar{f} > 0$, we have $M_f = M$, $\tilde{f} = f$, and since k^s_α is bounded below, f is bounded below. Moreover, since

$$f = 2^{-n} K^s_\alpha \bar{f}_n + K^s_\alpha(\Sigma_{l \neq n} 2^{-l} \bar{f}_l)$$

and $K^s_\alpha(\Sigma_{l \neq n} 2^{-l} \bar{f}_l)$ is bounded below (as $\Sigma_{l \neq n} 2^{-l} \bar{f}_l > 0$), it follows that $\int_0^{1/n} f \circ X_\sigma(\omega) d\sigma = \infty$ for all n.

We now turn to the proof of Theorem 1 (1)(c) for the

case $d = 1$. By definition $l \in H^{-\infty}(M)$ is in $H^{\alpha}(M)$ for some $\alpha > 0$ if there exists $f_1 \in H^{\alpha}(M)$ s.t. $l(\phi) = \int_M f_1 \phi d\lambda$ for $\phi \in C^{\infty}(M)$. The space $H^{\alpha}(M)$, by the way, can be considered as the dual of $H^{-\alpha}(M)$, if we let $l(l') = l'(l)$ for $l \in H^{\alpha}(M)$, $l' \in H^{-\alpha}(M)$. The common value equals $\int_M K^s_{-\alpha} f_1 \cdot K^s_{\alpha} l' d\lambda$.

If $d = 1$, any (M,g) is isometric to $S^1 = \{\alpha e^{2\pi i\beta}, \beta \in [0,1)\}$, endowed with the metric of R^2. Without loss of generality we may assume $\alpha = 1/2\pi$, so that $m(S^1) = 1$. Moreover, in this case $L_t \ll m$ for any Brownian motion on S^1. We denote by $(L\cdot,\cdot\cdot)$ the local time of the latter, which is the continuous version of dL_t/dm. Obviously, $L_t = L_t \circ \tilde{}$. As explained above, we have for $\alpha > 0$, that $\tilde{L}_t \in H^{\alpha}(S^1)$ iff $L(t,\cdot\cdot) \in H^{\alpha}(S^1)$, and in this case $\tilde{L}_t \in$ dual of $H^{-\alpha}(S^1)$ with $\tilde{L}_t(l') = \int_{S^1} \overline{K}_{\alpha} l' \cdot \overline{K}_{-\alpha} L(t,\cdot\cdot)dm$, $l' \in H^{-\alpha}(M)$, $\overline{K}_{\alpha}$ corresponding to Brownian motion on S^1 with $V = 0$.

For the proof of Theorem 1 (1)(c) for $d = 1$ we shall use the following lemma, which may be of independent interest.

LEMMA 1: Let L be the local time of Brownian motion on S^1 without drift. For $y \in S^1$, $r \in (0, \frac{1}{2})$ let $b(y,r)$ be the components on one side of y of the punctured balls $B(y,r) - \{y\}$ on S^1. Let $c > 0$, and let

$$F(t,y,r) = m\{z \in b(y,r);\ \frac{L(t,z) - L(t,y)}{\sqrt{r(y,z)}} > c\sqrt{L(t,y)}\},$$

$$t > 0$$

and

$$\Omega_{t,y} = \{\omega;\ \underline{\lim}_{r\to 0} - \frac{1}{\log r} \int_r^{1/2} \frac{F(t,y,\rho)}{\rho^2}\, d\rho > 0\}.$$

Then $P^x(\Omega_{t,y}) = 1$, all $x \in S^1$.

PROOF: Since Brownian motion on S^1 can be considered as Brownian motion on $\mathbf{R}^1$ (mod 1), it suffices to prove the following statement for Brownian motion on $\mathbf{R}^1$: For $z \in [-1/2, 1/2)$, $t > 0$ let $A(t,z) = \Sigma_{n=-\infty}^{\infty} L(t, z + n)$, where L denotes now the local time of Brownian motion on $\mathbf{R}^1$ and the sum is in fact a finite sum. For $r \in (0, 1/2)$, $t > 0$ let

$$G(t,r) = 1\{z \in (0,r);\ \frac{A(t,z) - A(t,0)}{\sqrt{z}} > c\sqrt{A(t,0)}\ \}.$$

Then for fixed $t > 0$, $x \in [-1/2, 1/2)$, P^x - a.e.

$$(*) \qquad \underline{\lim}_{r\to 0} - \frac{1}{\log r} \int_r^{1/2} \frac{G(t,\rho)}{\rho^2}\, d\rho > 0.$$

It obviously suffices to prove that (*) holds P^x - a.e. on the set $\{\tau_{\mathbf{Z}} < t\}$, where τ_A denotes now the first hitting time for the set A of Brownian motion on $\mathbf{R}^1$.

In the following fix also $N \in \mathbf{N}$, $N > 3$. As $P^x\{X_t \in \mathbf{Z}\} = 0$, it is sufficient to prove that (*) holds P^x - a.e. on

$$\Omega^* = \{\omega;\ \tau_{\mathbf{Z}} < t,\ 1/N < X_t - [X_t] < 1 - 1/N\}.$$

Let $\mathbf{Z}_N = \{\mathbf{Z} + 1/N\} \cup \{\mathbf{Z} - 1/N\}$, and denote by ζ_1, $\zeta_1 + \eta_1$, ζ_2, $\zeta_2 + \eta_2$, ... the successive hitting times for the sets $\mathbf{Z}$, $\mathbf{Z}_N$, $\mathbf{Z}$, $\mathbf{Z}_N$, ... For $k > 1$, let

$$\Omega(k) = \{\omega;\ \zeta_k + \eta_k < t,\ \zeta_{k+1} > t\}.$$

Obviously, $\Omega^* \subseteq \bigcup_{k\geq 1}\Omega(k)$. Moreover, on $\Omega(k)$, $\delta < X_s - [X_s] < 1 - \delta$ for $s \in [\zeta_k + \eta_k, t]$ and sufficiently small $\delta > 0$. It follows that on $\Omega(k)$, $A(t,z) = A(\zeta_k + \eta_k, z)$ for sufficiently small $|z|$, and hence $G(t,r) = G(\zeta_k + \eta_k, r)$ for sufficiently small r. It is therefore sufficient to prove for fixed $k \geq 1$, that P^x - a.e. on Ω,

$$\varliminf_{r\to 0} - \frac{1}{\log r}\int_r^{1/2} \frac{G(\zeta_k + \eta_k, \rho)}{\rho^2}\, d\rho > 0.$$

Introducing the independent processes

$$\{L_\nu(z),\ z \in [-1/2, 1/2)\}_{\nu\geq 1},\ L_\nu(z) = L(\eta_\nu, z + X_{\zeta_\nu}) \circ \theta_{\zeta_\nu},$$

we notice that $A(\zeta_k + \eta_k, z) \geq \Sigma_{\nu=1}^{k} L_\nu(z)$, and $A(\zeta_k + \eta_k, 0) = \Sigma_{\nu=1}^{k} L_\nu(0)$. Letting

$$H(r) = H(k,r) = 1\{z \in (0,r); \frac{\Sigma_{\nu=1}^{k}\{L_\nu(z) - L_\nu(0)\}}{\sqrt{z}} \geq c\sqrt{\Sigma_{\nu=1}^{k} L_\nu(0)\}}\},$$

$r \in (0, 1/2)$, we notice that $H(r) \leq G(\zeta_k + \eta_k, r)$. It is therefore sufficient to prove that for fixed $k \geq 1$, P^x - a.e. on Ω

$$(**) \qquad \varliminf_{r\to 0} - \frac{1}{\log r}\int_r^{1/2} \frac{H(\rho)}{\rho^2}\, d\rho > 0.$$

Now consider for fixed $k \geq 1$, the set $\mathscr{I}$ of indices $I = \{\varepsilon_1, \ldots, \varepsilon_k;\ n_1, \ldots, n_k\}$, where $\varepsilon_\nu = +1$ or -1, $n_1 = 0$ or

-1 if $x \in [-\frac{1}{2}, 0)$, $n_1 = 0$ or $+1$ if $x \in [0, \frac{1}{2})$, $n_{\nu+1} = n_\nu$ or $n_\nu + 1$ if $\varepsilon_\nu = +1$, $n_{\nu+1} = n_\nu$ or $n_\nu - 1$ if $\varepsilon_\nu = -1$. Let

$$\Omega_I = \{\omega; X_{\zeta_\nu} = n_\nu, X_{\zeta_\nu+\eta_\nu} = n_\nu + \varepsilon_\nu 1/N, \nu = 1,\ldots,k\},$$

and for $\varepsilon \in (0,1/N)$ let

$$\Omega_{I,\varepsilon} = \{\omega \in \Omega_I; (\forall \nu < k; \varepsilon_\nu = -1) (\exists s \in (\zeta_\nu, \zeta_\nu + \eta_\nu)) X_s$$

$$= n_\nu + \varepsilon\}.$$

Since $\Omega = \bigcup_{I \in \mathcal{I}} \Omega_I$ and $\Omega_{I,\varepsilon} \uparrow \Omega_I$ a.e. as $\varepsilon \downarrow 0$, it suffices to prove that for fixed k, I, ε, (**) holds P^x- a.e. on $\Omega_{I,\varepsilon}$.

Let $l = k + \#\{\nu < k, \varepsilon_\nu = -1\}$. We denote by $\xi_1,\ldots,\xi_{2l}$ the sequence obtained from the sequence n_1, $n_1 + \varepsilon_1 1/N, n_2,\ldots,n_k, n_k + \varepsilon_k/N$ by inserting $n_\nu + \varepsilon$, $n_\nu + \varepsilon$ between n_ν and $n_\nu + \varepsilon_\nu/N$, whenever $\varepsilon_\nu = -1$. Similarly let $m_1,\ldots,m_l$ denote the sequence obtained from the sequence $n_1,\ldots,n_k$ by inserting n_ν between n_ν and $n_{\nu+1}$, whenever $\varepsilon_\nu = -1$. We define the stopping times $\zeta_1', \zeta_1' + \eta_1'$, $\zeta_2', \zeta_2' + \eta_2', \ldots, \zeta_l', \zeta_l' + \eta_l'$ as the successive hitting times for the points $\xi_1,\ldots,\xi_{2l}$. If we let further $L_\nu'(z) = L(\eta_\nu', z + m_\nu) \circ \theta_{\zeta_\nu'}$, $z \in [-\frac{1}{2}, \frac{1}{2})$, $\nu = 1,\ldots,l$, and if $H'(r)$, $r \in (0, \frac{1}{2})$, is defined on Ω like $H(r)$, $r \in (0, \frac{1}{2})$, except that L_ν is replaced by L_ν' and k by l, then we have on $\Omega_{I,\varepsilon}$

$$\Sigma_{\nu=1}^{k} L_\nu(z) = \Sigma_{\nu=1}^{l} L_\nu'(z), \quad z \in [-\tfrac{1}{2}, \tfrac{1}{2}),$$

and therefore $H(r) = H'(r)$, $r \in (0, \frac{1}{2})$.

It is hence sufficent to show that for fixed $k \geq 1$, $I \in \mathscr{I}$, $\varepsilon \in (0,1/N)$, P^x - a.e. on Ω, (**) holds with H' in place of H. Letting for $r \in (0, \frac{1}{2})$

$$H^*(r) = 1\{z \in (0,r);\ \frac{L'_\nu(z) - L'_\nu(0)}{\sqrt{z}} > c\sqrt{L'_\nu(0)}\ ,$$

$$\text{all } \nu \leq l\},$$

we notice that $H^* \leq H'$. Thus it suffices to show that P^x - a.e. on Ω, (**) holds with H^* in place of H.

For this proof we notice first that the $\{L'_\nu(z),\ z \in [-\frac{1}{2}, \frac{1}{2})\}_{\nu=1,\ldots,l}$ are independent processes. We shall also use the following result of D. Ray (see e.g. [11]): For Brownian motion on $\mathbf{R}^1$ starting at 0, the process $\{2L(\tau_1, 1 - \xi),\ 0 \leq \xi \leq 1\}$ is the square of a two-dimensional Bessel process. It follows easily that for $a > 0$ the process $\{L(\tau_a, \xi),\ 0 \leq \xi \leq a\}$ has the same law as the process $\{\frac{1}{2} R^2_{a-\xi},\ 0 \leq \xi \leq a\}$, where $\{R_t,\ t \geq 0\}$ is a two-dimensional Bessel process starting at 0. In particular, for each $\nu \leq l$, $\{L'_\nu(z),\ z \in (0,\varepsilon)\}$ has the same distribution as $\{\frac{1}{2} R^2_{t_\nu(z)},\ z \in (0,\varepsilon)\}$, where each function t_ν is one of the functions $z \to \varepsilon - z$, $z \to 1/N + z$, $z \to 1/N - z$, the actual value depending on the sequence $\varepsilon_1,\ldots,\varepsilon_k$. Therefore, if $\{Y^{i\nu}_t,\ t \geq 0\}_{i=1,2,\nu \in \mathbf{N}}$ are independent one-dimensional Brownian motions starting at 0, on a space Ω^b, if $R^\nu_t = [(Y^{1\nu}_t)^2 + (Y^{2\nu}_t)^2]^{\frac{1}{2}}$, and if for $r \in (0,\varepsilon)$

$$H^b(r) = 1\{z \in (0,r);\ \frac{(R^\nu_{t_\nu(z)})^2 - (R^\nu_{t_\nu(0)})^2}{\sqrt{z}} > \sqrt{2}\, c\, R^\nu_{t_\nu(0)},$$

all $\nu < 1\}$,

then $\{H^*(r),\ r \in (0,\varepsilon)\}$ and $\{H^b(r),\ r \in (0,\varepsilon)\}$ have the same laws. It is therefore sufficient to show that a.e. on Ω^b

$$(+) \qquad \underline{\lim}_{r \to 0} - \frac{1}{\log r} \int_r^\varepsilon \frac{H^b(\rho)}{\rho^2}\, d\rho > 0.$$

Let now $S = \{(s_{i\nu});\ i = 1,2;\ \nu = 1,\dots,l\}$, with $s_{i\nu} =$ +1 or -1. For $s \in S$ let

$$\Omega_s^b = \{\omega^b;\ \operatorname{sgn} Y_{t_\nu}^{i\nu}(0) = s_{i\nu},\ i = 1,2,\ \nu = 1,\dots l\},$$

and for $s \in S$, $c' > \tfrac{1}{2}\sqrt{2}\, c$ consider the random function

$$H_s^b(r) = l\{z \in (0,r);\ s_{i\nu} \cdot \frac{Y_{t_\nu}^{i\nu}(z) - Y_{t_\nu}^{i\nu}(0)}{\sqrt{z}} > c'$$

$$\text{all } i = 1,2,\ \nu < 1\},$$

$r \in (0,\varepsilon)$, on Ω^b. We have for each $s \in S$ that on Ω_s^b, $H_s^b(r) < H^b(r)$ for sufficiently small r (depending on ω^b). As $\Omega^b \subseteq \bigcup_{s \in S} \Omega_s^b$ a.e., it is therefore sufficient to show that for fixed $s \in S$, (+) holds a.e. on Ω^b, if H^b is replaced by H_s^b, and this can be proved as follows.

Notice that the law of $\{Y_{t_\nu}^{i\nu}(z) - Y_{t_\nu}^{i\nu}(0),\ z \in (0,\varepsilon),\ i = 1,2,\ \nu = 1,\dots,l\}$ is the same as the law of $\{Y_z^{i\nu},\ z \in (0,\varepsilon),\ i = 1,2,\ \nu = 1,\dots,l\}$. Letting

$$K(r) = l\{z \in (0,r);\ s_{i\nu} \frac{Y_z^{i\nu}}{\sqrt{z}} > c' \text{ all } i = 1,2,\ \nu < 1\},$$

$$r > 0,$$

we conclude that the laws of $\{H_s^b(r),\ r \in (0,\varepsilon)\}$ and $\{K(r),\ r \in (0,\varepsilon)\}$ coincide. It remains to show that (+) holds a.e. if H^b is replaced by K. Now for all $h > 0$, the laws of $\{K(hr)/hr,\ r > 0\}$ and $\{K(r)/r,\ r > 0\}$ are the same, and as K increases and $0 < K(r)/r < 1$, the ergodic theorem gives

$$\lim_{r\to 0} -\frac{1}{\log r}\int_r^{\varepsilon}\frac{K(\rho)}{\rho^2}\,d\rho = \left(\frac{1}{\sqrt{2\pi}}\int_{c'}^{\infty} e^{-u^2/2}du\right)^{21} \text{ a.e.}$$

Proof of Theorem 1 (1)(c) for d = 1:

We have to show that on a shift invariant set Ω_0 of full measure, $L(t,\cdot\cdot) \notin H^{1/2}(S^1)$, $t > 0$. In view of Girsanov's theorem, it suffices to show this if $V = 0$. Now, we have by an easy modification of proposition 4 of V, 3.5 in [9], that for $\alpha \in (0,1)$, $f \in H^{\alpha}(S^1)$ iff $f \in L^2(S^1)$ and

$$\int_{S^1}\int_{S^1}\frac{[f(y)-f(z)]^2}{[r(y,z)]^{1+2\alpha}}\,dm(y)dm(z) < \infty.$$

It is therefore sufficient to show that for the set

$$\Omega_0 = \{\omega;\ \int_{S^1}\int_{S^1}\frac{[L(t,y)-L(t,z)]^2}{[r(y,z)]^2}\circ\theta_s\,dm(y)dm(z) = \infty,$$

$$\text{all } s > 0,\ t > 0\}$$

we have $P^x(\Omega_0) = 1$, $x \in S^1$. This will be done in several steps. Let $y \in S^1$, $t > 0$. We have for $\omega \in \Omega_{t,y}$ of Lemma 1 that

$$\int_{S^1}\frac{[L(t,z)-L(t,y)]^2}{[r(y,z)]^2}\,dm(z) > c^2 L(t,y)\ \times$$

$$\int_{b(y,\varepsilon)-b(y,r)} \frac{1}{r(y,z)} \times$$

$$\chi_{\{(L(t,z)-L(t,y)\geq c\sqrt{L(t,y)r(y,z)}\}}(z)dm(z)$$

$$= c^2L(t,y) \int_r^\varepsilon 1/\rho \, d_\rho F(t,y,\rho)$$

$$= c^2L(t,y) \left\{\frac{F(t,y,\rho)}{\rho}\Big|_{\rho = r}^{\rho = \varepsilon} + \int_r^\varepsilon \frac{F(t,y,\rho)}{\rho^2} d\rho\right\}$$

$$\geq -c^2L(t,y)c'(\omega)\log r$$

for sufficiently small r, where $c'(\omega) > 0$. Letting for $y \in S^1$, $t > 0$,

$$\Omega^1_{t,y} = \{\omega;\ \int_{S^1} \frac{[L(t,y) - L(t,z)]^2}{[r(y,z)]^2} dm(z) = \infty\}$$

and denoting by τ_y the first hitting time for y, we conclude that

$$\{\tau_y < t\} \cap \Omega_{t,y} \subseteq \Omega^1_{t,y} \subseteq \{\tau_y < t\}$$

and by Lemma 1 that $P^x\{\Omega^1_{t,y} \,\Delta\, \{\tau_y < t\}\} = 0$, $x \in S^1$. It follows by Fubini's theorem that for $t > 0$

$$P^x\{\omega;\ \int_{S^1} \frac{[L(t,y) - L(t,z)]^2}{[r(y,z)]^2} dm(z) = \infty$$

$$\text{a.a. } y \in X.[0,t]\} = 1,\ x \in S^1,$$

and then by the Markov property of Brownian motion that for $t > 0$, $s > 0$

$$P^x\{\omega;\ \int_{S^1} \frac{[L(t,y) - L(t,z)]^2 \cdot \theta_s}{[r(y,z)]^2}\, dm(z) = \infty$$

$$\text{a.a. } y \in X.[s,s+t]\} = 1,\ x \in S^1.$$

If we let

$$\Omega_1 = \{\omega;\ \text{all } s > 0,\ t > 0 \text{ rat. :}$$

$$\int_{S^1} \frac{[L(t,y) - L(t,z)]^2 \circ \theta_s}{[r(y,z)]^2}\, dm(z) = \infty$$

$$\text{a.a. } y \in X.[s,s+t]\}$$

$$\Omega_2 = \{\omega;\ \text{all } s > 0,\ t > 0 \text{ :}$$

$$\int_{S^1} \frac{[L(t,y) - L(t,z)]^2 \circ \theta_s}{[r(y,z)]^2}\, dm(z) = \infty$$

$$\text{a.a. } y \in X.[s,s+t]\},$$

then obviously, $P^x(\Omega_1) = 1$, $x \in S^1$, and $\Omega_2 \subseteq \Omega_1$. But we also have $\Omega_1 \subseteq \Omega_2$: For $\omega \in \Omega_1$, $s > 0$, $t > 0$, there exists $E_\omega \subseteq X.[s,s+t]$ s.t. $m(X.[s,s+t] - E_\omega) = 0$ and such that for $y \in E_\omega$ and s', t' rational with $s' < s$, $s + t < s' + t'$,

$$\int_{S^1} \frac{[L(t',y) - L(t',z)]^2 \circ \theta_{s'}}{[r(y,z)]^2}\, dm(z) = \infty,$$

Now for $y \in S^1 - \{X_s, X_{s+t}\}$, there exists $\delta > 0$ and s', t' rational with $s' < s$, $s + t < s' + t'$ such that $X_\sigma \notin B(y,\delta)$ for $\sigma \in [s',s] \cup [s+t, s'+t']$, and hence $L(t',z) \circ \theta_{s'} =$

$L(t,z) \circ \theta_s$ for $z \in B(y,\delta)$. It follows that for $y \in E_\omega - \{X_s, X_{s+t}\}$, we have

$$\int_{S^1} \frac{[L(t,y) - L(t,z)]^2 \circ \theta_s}{[r(y,z)]^2} \, dm(z) = \infty,$$

which proves $\Omega_1 \subseteq \Omega_2$. It follows now easily from $P^x(\Omega_2) = 1$, $x \in S^1$, that $P^x(\Omega_0) = 1$, $x \in S^1$. For the proof of Theorem 3 we need two lemmas.

LEMMA 2: For $f \in H^2(M)$, $\int_M |\text{grad } f|^2 d\lambda = -2 \int_M f \cdot Lf \, d\lambda$

PROOF: Since $|\text{grad } \cdot| : H^1(M) \to L^2(M)$, $L : H^2(M) \to L^2(M)$ are continuous and $C^\infty(M)$ is dense in $H^2(M)$, it is sufficient to prove the formula for $f \in C^\infty(M)$. From $L'\phi = 0$ and the divergence theorem we obtain for $f \in C^\infty(M)$

$$\begin{aligned}\int_M |\text{grad } f|^2 d\lambda &= - \int_M f \cdot \Delta f d\lambda + \tfrac{1}{2} \int_M \Delta(f^2) \cdot \phi dm \\ &= -2 \int_M f \cdot Lf d\lambda + \tfrac{1}{2} \int \Delta(f^2) \cdot \phi dm + \int_M V(f^2) \cdot \phi dm \\ &= -2 \int_M f \cdot Lf d\lambda + \int_M f^2 \cdot L'\phi dm \\ &= -2 \int_M f \cdot Lf d\lambda.\end{aligned}$$

REMARK: Lemma 2 implies that for $f \in L^2(M)$

$$\int_M |\text{grad } Gf|^2 d\lambda = 2 \int_M f \cdot Gf d\lambda.$$

By Theorem (3.2)(a) in [1], the operators in the following chain are Hilbertian space isomorphisms:

$$L^2_\lambda(M) \xrightarrow[G]{} H^2_\lambda(M) \xrightarrow[L^s]{} L^2_\lambda(M) \xrightarrow[G^*]{} H^2_\lambda(M).$$

It follows that $G + G^* = -2\, G^*L^sG$ is a Hilbertian space isomorphism $G + G^* : L^2_\lambda(M) \to H^2_\lambda(M)$. Its inverse $(G + G^*)^{-1} : H^2_\lambda(M) \to L^2_\lambda(M)$ is

$$(G + G^*)^{-1} = -\tfrac{1}{2}\, L(L^s)^{-1}L^*$$

$$= -\tfrac{1}{2}\,(L^s + V_1)(L^s)^{-1}(L^s - V_1)$$

$$= -\tfrac{1}{2}\,(L^s - V_1(L^s)^{-1}V_1).$$

LEMMA 3: <u>The operator</u> $(G + G^*)^{1/2}$, <u>defined on the Hilbert space</u> $L^2_\lambda(M,\lambda)$, <u>is a Hilbertian space isomorphism</u> $(G + G^*)^{1/2} : L^2_\lambda(M) \to H^1_\lambda(M)$.

PROOF: The argument is essentially the interpolation argument given in [1] for K^s_α. As $G + G^*$ is a Hilbertian space isomorphism $G + G^* : L^2_\lambda(M) \to H^2_\lambda(M)$, we may consider $H^2_\lambda(M)$ as the underlying Hilbertian space of the Hilbert space $\overline{H}^2_\lambda(M)$, defined by $\overline{H}^2_\lambda(M) = (G + G^*)L^2_\lambda(M)$, endowed with the inner product

$$((G + G^*)f_1,\ (G + G^*)f_2)_{\overline{H}^2_\lambda(M)} = \int_M f_1 f_2 d\lambda.$$

It is well-known that $H^1_\lambda(M)$ is the underlying Hilbertian space of the Hilbert space $Q_{1/2}\,(\overline{H}^2_\lambda(M),\ L^2_\lambda(M,\lambda))$, where Q denotes quadratic interpolation of Hilbert spaces. The latter space is by definition the completion of $\overline{H}^2_\lambda(M)$, realized in $L_\lambda^2(M,\lambda)$, with respect to the norm $\|f\|_Q = \|(G + G^*)^{1/2} f\|_{\overline{H}_\lambda^2(M)}$. It suffices to identify this space with $\overline{H}_\lambda^1(M) = (G^\lambda + G^*)^{1/2} L_\lambda^2(M)$, endowed with the inner product

$$((G + G^*)^{1/2} f_1, (G + G^*)^{1/2} f_2)_{\overline{H}_\lambda^1(M)} = \int_M f_1 f_2 d\lambda.$$

Notice that $\overline{H}_\lambda^2(M) \subseteq \overline{H}_\lambda^1(M)$ and that $\overline{H}_\lambda^1(M)$ is complete. Moreover, $H_\lambda^2(M)$ is dense in $\overline{H}_\lambda^1(M)$, because denseness of $H_\lambda^2(M)$ in $L_\lambda^2(M)$ implies denseness of $(G + G^*)^{1/2} H_\lambda^2(M)$ in $\overline{H}_\lambda^1(M)$ and because $(G + G^*)^{1/2} H_\lambda^2(M) \subseteq H_\lambda^2(M)$. It follows that $\overline{H}_\lambda^1(M)$ is the completion of $\overline{H}_\lambda^2(M)$, realized in $L_\lambda^2(M,\lambda)$, with respect to the norm $\| \ \|_{\overline{H}_\lambda^1(M)}$. Now notice that for $f \in \overline{H}_\lambda^2(M)$, $\|f\|_{\overline{H}_\lambda^1(M)} = \|f\|_Q$. This implies $\overline{H}_\lambda^1(M) = Q_{1/2}(\overline{H}_\lambda^2(M), L_\lambda^2(M,\lambda))$.

<u>Proof of Theorem 3:</u>

We have by Lemma 3

$$(G + G^*)^{1/2} B = \{(G + G^*)^{1/2} f; \int_M f^2 d\lambda < 1\}$$

$$= \{f \in H_\lambda^1(M); \int_M |(G + G^*)^{-1/2} f|^2 d\lambda < 1\}.$$

Moreover, by Lemma 2 we have for $f \in H_\lambda^2(M)$

$$\int_M |(G + G^*)^{-1/2} f|^2 d\lambda$$

$$= \int_M f \cdot (G + G^*)^{-1} f d\lambda$$

$$= -1/2 \int_M f \cdot (L^s - V_1 (L^s)^{-1} V_1) f d\lambda$$

$$= \tfrac{1}{4} \int_M |\text{grad } f|^2 d\lambda + \tfrac{1}{2} \int_M f \cdot V_1 (L^s)^{-1} V_1 f d\lambda.$$

As div $(\phi V_1) = 0$, the last integral equals $-\tfrac{1}{2} \int_M V_1 f \cdot (L^s)^{-1} V_1 f d\lambda$ which equals $\tfrac{1}{4} \int_M |\text{grad}(L^s)^{-1} V_1 f|^2 d\lambda$ by the remark following Lemma 2. By approximation we obtain for $f \in H^1_\lambda(M)$

$$\int_M |(G + G^*)^{-\frac{1}{2}} f|^2 d\lambda = \tfrac{1}{4} \int_M |\text{grad } f|^2 d\lambda$$

$$+ \tfrac{1}{4} \int_M |\text{grad}(L^s)^{-1} V_1 f|^2 d\lambda.$$

We conclude this paper with a short note. It has been pointed out to us by Mr. Hsu Pei that the short proof of theorem (3.8) In [2] contains a slip. We suggest that the reader make the following changes in that proof:

(1) Omit the second part of (3.10) and add instead the following lines

"Notice that for an eigenfunction ϕ with eigenvalue $\lambda > 0$ for arbitrary $\varepsilon \in (0, \tfrac{1}{2})$

$\|\text{grad } G\phi\|_\infty = \|\text{grad } G_{\frac{1}{2}+\varepsilon}(G_{\frac{1}{2}-\varepsilon}\phi\|_\infty \leq C_\varepsilon \|G_{\frac{1}{2}-\varepsilon}\phi\|_\infty$

$\leq \lambda^{(-\frac{1}{2}-\varepsilon)} \|\phi\|_\infty$"

after the first part of (3.10).

(2) Continue line -5 on p. 109 by

"for $\alpha' > \alpha$ with c depending on $\alpha' - \alpha$".

(3) In the remainder of the proof

replace α by α' in $M_t^{n,\alpha}$, $M^\alpha(t,\omega)$, ϕ_n^α, H_0^α, $\lambda_n^{-(\alpha+1)/2}$.

References

[1] J. R. Baxter and G. A. Brosamler, Recurrence of Brownian Motions on Compact Manifolds, Colloque en l'Honneur de Laurent Schwartz, Astérisque 132, 15-46 (1985).

[2] G. A. Brosamler, Laws of the Iterated Logarithm for Brownian Motions on Compact Manifolds, Z. Wahrscheinlichkeitstheorie verw. Gebiete 65, 99-114 (1983).

[3] Z. Ciesielski and S. J. Taylor, First Passage Times and Sojourn Times for Brownian Motion in Space and the Exact Hausdorff Measure of the Sample Path, Trans. Amer. Math. Soc. 103, 434-450 (1962).

[4] L. Elie, Equivalent de la densité d'une diffusion en temps petits. Cas des points process, Astérisque 84-85, 55-71 (1981).

[5] J. Kuelbs and R. LePage, The Law of the Iterated Logarithm for Brownian Motion in a Banach Space, Trans. Amer. Math. Soc. 185, 253-264 (1973).

[6] J. Kuelbs and W. Philipp, Almost Sure Invariance Principles for Partial Sums of Mixing B-Valued Rnadom Variables, Ann. Prob. 8, 1003-1036 (1980).

[7] H. P. McKean, Stochastic Integrals, Academic Press, New York (1969).

[8] E. Nelson, The Adjoint Markoff Process, Duke Math. J. 25, 671-690 (1958).

[9] E. M. Stein, Singular Integrals and Differentiability Properties of Functions, Princeton Univeristy Press, Princeton (1970).

[10] S. J. Taylor, The Exact Hausdorff Measure of the Sample Path for Planar Brownian Motion, Proc. Cambridge Phil. Soc. 60, 253-258 (1964).

[11] D. Williams, Markov Properties of Brownian Local Time, Bull. Amer. Math. Soc. 75, 1035-1036 (1969).

Gunnar A. Brosamler
Fachbereich Mathematik
Universitaet des Saarlandes
Saarbruecken, West Germany

and

Department of Mathematics
The University of British Columbia
Vancouver, B. C., Canada

Seminar on Stochastic Processes, 1985
Birkhäuser, Boston, 1986

CORRECTION TO: TOPICS IN ENERGY AND POTENTIAL THEORY

(Seminar on Stochastic Processes, 1982)

by

Joseph Glover

R. K. Getoor pointed out to us that the proof of Lemma (3.1) is not valid if b is ∞ or $-\infty$, so we give a new proof of (3.1) here which allows us to prove Theorem (3.2). We assume the reader is familiar with the notation in the article.

(3.1) LEMMA Assume (E). Let n and v be signed measures with $(|v|,|v|) < \infty$ and $(|n|,|n|) < \infty$. If v does not charge any polar set, then $|(n,v) + (v,n)| \leq 2\|n\| \, \|v\|$.

PROOF Case 1: assume $(|n|,|v|) + (|v|,|n|) < \infty$ and $(n,n) > 0$. Set $b = -((n,v) + (v,n))/2(n,n)$. Then $0 \leq (bn + v, bn + v)$ implies $((n,v) + (v,n))^2 \leq 4(v,v)(n,n)$.

Case 2: assume $(n,n) = 0$. Let $G = \{Un + n\hat{U} > 0\}$, and choose a finite positive measure m carried by G so that $(m + |n|, m + |n|) < \infty$. Then for every $c > 0$, $0 \leq (cn - m, cn - m) = (m,m) - c[(m,n) + (n,m)]$. This forces $(m,n) + (n,m)$ to be zero, so G must be empty. A similar argument applies to show $H = \{Un + n\hat{U} < 0\}$ is empty. Therefore, $|(n,v) + (v,n)| = 0$.

Case 3: assume $(|n|,|v|) + (|v|,|n|) = \infty$ and $(n,n) > 0$. Since v does not charge any polar sets, $|v|\{U|n| + |n|\hat{U} = \infty\} = 0$. Choose $f_k > 0$ increasing to 1 so that if $v_k = f_k v$, then $A_k = (|n|,|v_k|) + (|v_k|,|n|) < \infty$. Then $B_k = (|n| - |v_k|,|n| - |v_k|) = (|n|,|n|) + (|v_k|,|v_k|) - A_k$. Since $\lim_{k\to\infty}(|v_k|,|v_k|) = (|v|,|v|) < \infty$, we have $B_k < 0$ for some k, contradicting (E). Therefore, this case cannot occur. Q.E.D.

This lemma is enough to complete the proof of Theorem (3.2): (E) implies (M). It is used in the next to last line of the proof: $\mu U\gamma < 2\|\mu\|\ \|\gamma\| < \infty$. Note that γ does not charge any polar set by construction. Once we know (M) holds, we know that if a measure μ charges a polar set, then $\mu\{U\mu = \infty\} > 0$. It follows that the energy inequality $|(n,v) + (v,n)| < 2\|n\|\ \|v\|$ holds **without** assuming that v does not charge a polar set.

Joseph Glover

Department of Mathematics

University of Florida

Gainesville, Florida 32611

Progress in Probability and Statistics

1 CINLAR/CHUNG/GETOOR. Seminar on Stochastic Processes, 1981
ISBN 3-7643-3072-4
2 KESTEN. Percolation Theory for Mathematicians
ISBN 3-7643-3107-0
3 ASMUSSEN/HERING. Branching Processes
ISBN 3-7643-3122-4
4 CHUNG/WILLIAMS. Introduction to Stochastic Integration
ISBN 3-7643-3117-8
5 CINLAR/CHUNG/GETOOR. Seminar on Stochastic Processes, 1982
ISBN 3-7643-3131-3
6 BLOOMFIELD/STEIGER. Least Absolute Deviation
ISBN 3-7643-3157-7
7 CINLAR/CHUNG/GETOOR. Seminar on Stochastic Processes, 1983
ISBN 3-7643-3293-X
8 BOUGEROL/LACROIX. Products of Random Matrices with Application to Schrödinger Operator
ISBN 3-7643-3324-3
9 CINLAR/CHUNG/GETOOR. Seminar on Stochastic Processes, 1984
ISBN 3-7643-3327-8
10 KIFER. Ergodic Theory of Random Transformations
ISBN 3-7643-3319-7
11 EBERLEIN/TAQQU. Dependence in Probability and Statistics
ISBN 3-7643-3323-5
12 CINLAR/CHUNG/GETOOR. Seminar on Stochastic Processes, 1985
ISBN 3-7643-3331-6